64798

2/16/07

Another Day in the Frontal Lobe

Another Day in the

FRONTAL LOBE

A Brain Surgeon Exposes Life on the Inside

KATRINA FIRLIK

RANDOM HOUSE / NEW YORK

Published in the United States by Random House, an imprint of The Random House Publishing Group, a division of Random House, Inc., New York.

RANDOM HOUSE and colophon are registered trademarks of Random House, Inc.

Grateful acknowledgment is made to Alfred A. Knopf, a division of Random House, Inc., for permission to reprint "What the Doctor Said" from *All of Us: The Collected Poems* by Raymond Carver, copyright © 1996 by Tess Gallagher, Introduction copyright © 1996 by Tess Gallagher, Editor's Preface, Commentary, and Notes copyright © 1996 by William J. Stull. Reprinted by permission of Alfred A. Knopf, a division of Random House, Inc.

LIBRARY OF CONGRESS CATALOGING-IN-PUBLICATION DATA

Firlik, Katrina.
  Another day in the frontal lobe: a brain surgeon exposes life on the inside / Katrina Firlik.
    p. cm.
  ISBN 1-4000-6320-5
  1. Firlik, Katrina. 2. Neurosurgeons—United States—Biography. 3. Nervous system—Surgery—United States. 4. Brain—Surgery—United States. I. Title.
  RD592.9.F57A3 2006   617.4'8092—dc22   2005055260
            [B]

Printed in the United States of America on acid-free paper

www.atrandom.com

9 8 7 6 5 4 3 2 1

FIRST EDITION

Book design by Casey Hampton

For Andy

Contents

Another Day in the Frontal Lobe

Scientist and Mechanic

The brain is soft. Some of my colleagues compare it to toothpaste, but that's not quite right. It doesn't spread like toothpaste. It doesn't adhere to your fingers the way toothpaste does. Tofu—the soft variety, if you know tofu—may be a more accurate comparison. If you cut out a sizable cube of brain it retains its shape, more or less, although not quite as well as tofu. Damaged or swollen brain, on the other hand, is softer. Under pressure, it will readily express itself out of a hole in the skull made by a high-speed surgical drill. Perhaps the toothpaste analogy is more appropriate under these circumstances.

The issue of brain texture is on my mind all the time. Why? I am a neurosurgeon. The brain is my business. Although I acknowledge that the human brain is a refined, complex, and mysterious system, I often need to regard it as a soft object inhabiting the bony confines of a hard skull. Many of the brains I encounter have been pushed around by tumors, blood clots, infections, or strokes that have swollen out of control. Some have been invaded by bullets, nails, or even maggots. I see brains at their most vulnerable. However, whereas other brain specialists, like neurologists and psychiatrists, examine brain images and pontificate from outside of the cranium, neurosurgeons boast the additional manual relationship with our most complex of organs. We are part scientist, part mechanic.

The scientist in me revels in the ethereal manifestations of the brain: the mind, consciousness, memory, language. The mechanic in me is satisfied by the clear fluid that rushes out of the end of a tube I insert into a patient's brain to relieve excessive pressure. In everyday surgical practice, the science may take a backseat to the handiwork, and that's okay. If you have an expanding blood clot in your head, you want a skilled brain mechanic, and preferably a swift one. You don't care if your surgeon published a paper in *Science* or *Nature*.

I'll give you an example of a most straightforward and manual case. I was paged to the emergency room a few years ago during my training and received the following brief report over the phone: "carpenter coming in with a nail stuck in the left frontal region of his head . . . neurologically intact." What is going through my mind at this point? Do I hark back to my studies of frontal lobe circuitry and mull over the complex neural networks involved in language and memory? No. I'm thinking concrete, surgical thoughts: nails are sharp; the brain is full of blood vessels; the nail may have snagged a vessel on the way in. These thoughts are instantaneous, of course. I spell out the simple logic here purely for effect.

What I encountered in the ER was a young man, in his thirties, sitting up on an emergency room gurney. Perfectly awake and alert, arms crossed in repose and still in his construction boots, he smiled nervously when I walked in. Was he the right patient? He looked too good.

He was the right one. The carpenter explained that he and his friend were both on ladders along the side of a house. His friend was working a few rungs above. They were driving heavy-duty nails into the siding with automatic nail guns. His friend's hand slipped upon firing in one of the nails, and the nail entered the left frontal region of my patient's head below. For the first few moments after impact, the carpenter doubted what had happened. Although he noticed a stinging sensation within a split second of his friend's slip of the hand, and heard the loud expletive coming from the same direction, there was no trickle of blood and he felt nothing unusual as his fingers frantically searched the top of his head. He wasn't sure if it went in. His friend knew otherwise.

Upon close inspection of his scalp, past his short crew cut, I could see the flat silver head of the nail, not quite flush with the scalp, but a bit deeper. Apart from the nail, he looked great. I performed a quick five-minute neurological exam and found nothing wrong. I sent him down the hall for a CT scan. The nail entered his brain perfectly perpendicular to the surface of the skull. It had been driven a good two inches into his left frontal lobe. Luckily, it didn't snag any sizable blood vessels along the way. There was no evidence of bleeding within the brain. Unlike the more common gunshot wounds we see, this was a respectably neat and clean penetrating injury.

At this point, my biggest fear—bleeding in the brain from entry of the nail—had been put to rest. Now, do I take a breath and mull over any complex scientific issues at this point? Am I exercising my formidable brainpower as a brain surgeon? When people say, "it doesn't take a brain surgeon," they refer to the assumption that we are the smartest ones around. Have I demonstrated this superior intelligence so far? Again, my thoughts return to the practical and concrete. We need to get the nail out of this guy's head. It didn't cause any bleeding on the way in. We need to avoid bleeding on the way out.

I walked out to the waiting room. His wife was there and so was his friend, who was pale and despondent, looking down at the floor. I tried to cheer them up a bit. Yes, the nail entered his brain, but his brain function, as far as we could tell, was normal and the nail caused no bleeding. Without looking up, the friend opened his hand and offered me a large silver nail that had been warming in his palm, the same type embedded in my patient's head. "I don't know . . . it might help you guys to have one of these . . . so you know what you're dealing with." I hadn't been able to tell from the scan that the nail had two copper-colored barbs sticking out from the shaft at acute angles. I'm not a carpenter, but I figured that the purpose of the barbs was to ensure a strong hold. I thanked him and pocketed the nail in my white coat. On my way back to the ER, I ran my fingers over the pointy barbs and thought about the issue of bleeding again. Avoiding and controlling bleeding are elementary and pervasive themes in surgery—not quite the stuff of rocket science, but critical nonetheless.

After calling on the appropriate team, including the supervising neurosurgeon and anesthesiologist, I took him to the OR, shaved a small patch of hair around the nail head, and made a short linear incision in his scalp, down to the skull. There are no how-to entries in our textbooks regarding removing nails from heads, so we improvised using common sense. We drilled out a disc of frontal bone from his skull, with the nail head at the center of the disc. Slowly, we lifted this piece of bone up away from the surrounding skull, bringing the firmly embedded barbed nail with it. Although we could see a small jagged tear in the covering of the brain and a puncture wound on the surface of the brain itself, there was no blood oozing from the hole, and we considered ourselves lucky. ("Better lucky than good" is a favorite slogan among surgeons.)

Then, using large tools fit more for our patient's line of work, we clipped off the barbs and pounded the nail through the disc of skull, backward. After soaking the bone in an antibiotic solution, we neatly plated it back in place with miniature titanium plates and screws and sewed his scalp back together. Actually, rather than suture, we used surgical staples from a staple gun to close the final layer of his scalp, unaware, at the time, of the subtle irony in that move. Within less than twenty-four hours, the patient was on his way home, joking the entire length of the hall with the friend who nailed him in the head.

When I recounted this story to my family and friends after dinner one night, they all nagged me with the same question: "How could he be normal? This went into his *brain*." Finally, here's where the scientist in me gets to pontificate a bit, settling into a fast-paced question-answer session in the comfort of my own home with a captive audience. I am not just a mechanic, after all, and the brain is not just tofu.

How could he be normal? First of all, his brain function was considered normal based on our typical bedside examination, which is, admittedly, a bit coarse. His speech was fluent. He answered simple questions appropriately. I asked him to remember three objects over a five-minute time span, and he did. His pupils reacted when I flashed a light in his eyes and his eyes moved symmetrically. He had no drooping of his face. The strength in his arms and legs was normal and so was his sensation. His reflexes were fine. He was capable of rapid and coordinated hand movements. In other words, his five-minute neurological examination was perfectly satisfactory.

But the frontal lobes harbor quite sophisticated functions, more sophisticated than the relatively simple ones I tested. The frontal lobes make up the largest section of the brain and are the most recently evolved. Compare the forehead of an ape to the forehead of a human. One slopes, the other bulges. We can thank, or blame, our frontal lobes for much of what we consider to be our personality and intelligence. Damage to the frontal lobes can be subtle, including changes in insight, mood, and higher-level judgment ("executive function," in the professional lingo). I'm not going to detect such changes in the ER during my five-minute exam before he is whisked off to the CT scanner. I'm just the neurosurgeon here. We would need to consult a neuropsychologist to help us evaluate these more complex brain functions.

"So why didn't you send this poor guy for more sophisticated testing?" my dinner audience asks in a confused and mildly accusatory tone. Why did I simply proclaim him "fine" and send him on his way? I explain that the foreign object was a nail, not a jackhammer. A relatively minuscule portion of brain was violated. The large frontal lobes, in particular, can be quite forgiving, especially when only one side is involved. It's not unusual to see a frontal lobe tumor, for example, grow to impressive citrus fruit proportions before the patient even detects a problem. In fact, the patient often does not detect a problem at all. It is frequently a spouse or friend who insists on the doctor appointment, explaining: "He's just not right, but I don't know what it is."

There is a redundancy and resilience to certain brain functions. What is compromised in one portion can sometimes be compensated for in another. (A remarkable ability referred to as "plasticity.") Even if the brain doesn't compensate directly, the patient often can cope indirectly, without even realizing it. If a person develops minor difficulty with memory, for example, he may start to write more things down, thereby maintaining the otherwise seamless flow of his existence. There are limits, though, to the power of plasticity. Damage to a single frontal lobe is frequently well tolerated (the opposite frontal lobe can compensate to some degree), whereas damage to both sides is often irreversibly devastating.

Getting back to our carpenter, we were confident that the very narrow swath of injured brain in only one frontal lobe would be inconsequential. Even if a faint cognitive deficit *could* be identified with detailed and time-consuming neuropsychological testing, would the patient really care? Would he, or anyone else, even notice the problem? Would his life as carpenter, husband, or friend be affected? Doubtful. On a more cold-blooded and practical note, would the patient or the hospital be willing to pay for these tests? His insurance would certainly balk at the cost and question the necessity. Besides, given my confidence in the resilience of his frontal lobes, my biggest concern was not sluggish thought but sluggish carpentry. What if he gives up the automatic nail gun altogether?

And with this final thought, the mechanic in me reclaims the front seat, as the scientist heads again to the back.

It doesn't necessarily take a brain surgeon to think like a brain surgeon, especially when it comes to the fundamentals. Consider this elementary notion: there is a limited amount of room inside the skull. Another central truth, directly related to the first, is: the brain is not the only thing inside the skull. The brain, in fact, makes up about 80 percent of the intracranial contents. The other 20 percent is split about evenly in volume between blood and cerebrospinal fluid. Once you master these central tenets, a good deal of seemingly complex neurosurgical decision-making becomes transparent.

Neurosurgeons learn to care just as much about the 20 percent as they do about the 80 percent, even though everyone else is blinded by the mystique of that 80 percent. I get plenty of wide-eyed questions about the brain, but no one ever cares to ask about the cerebrospinal fluid, a real nonissue as far as the public is concerned. Neurosurgeons care about the cerebrospinal fluid because if there's too much of it, the brain could be rendered next to useless.

In learning to think like a neurosurgeon, you have to take these thoughts one step further: given the rigid, fixed-volume container of the skull, and the 80/10/10 balance of its contents, what can be done if the equation is disrupted? This tips us more into the realm of mechanic than scientist.

Consider what would happen if you were punched in the eye. The area around the eye becomes swollen and is free to swell as much as it needs to. Aside from the social and cosmetic downsides of having a puffy, swollen eyelid and face, the swelling itself is usually not dangerous. It's not constrained. It goes down after several days, and the skin and underlying soft tissues recover nicely.

A swollen brain is another matter. There's not much room for it to swell. Swelling within a fixed container leads to elevated pressure, and unchecked pressure can lead to a cascade of events—namely a last-ditch shifting of delicate intracranial contents—that can be fatal. So as neurosurgeons, we do whatever we can to maintain a normal pressure within the skull when things go awry, such as in a serious head injury.

Although this is "brain surgery," the options we have for treating high pressures within the head are relatively simplistic and mechanistic: drain off cerebrospinal fluid from within the skull, shrink the brain tissue itself with a temporary dehydrating agent, or constrict the blood vessels in the brain via hyperventilation (although this one can be dangerous in situations when the brain needs all the blood flow it can get). If these options fail, there are more extreme measures, as a last resort: remove a portion of relatively "unimportant" brain tissue to create more room, or remove a section of skull to allow the brain to continue to swell. The decision as to which of these extreme measures you choose is largely a matter of who your mentor was and what he or she preferred to do.

As an example, here is how I handled such a decision recently. It was a warm sunny Sunday afternoon and I was sitting at an outdoor table at an Italian restaurant with my husband. We were eating salad and waiting for our Margherita pizza, content in our idle people-watching and discussing what we wanted to do after lunch. Our contentment was interrupted by my pager. It was one of the medical interns at the hospital: "Dr. Firlik, we need you to see a patient in the ICU, soon as you can. He had a stroke a few days ago and now he's herniating."

"Herniating" refers to the end-stage shifting of the brain in response to increased pressure. The word puts everyone into crisis mode, including my husband who, overhearing my conversation, had our pizza boxed up so he could eat it while sitting in the passenger seat of my car as I sped down the highway. He is also a neurosurgeon but, because of his passion for innovation and entrepreneurship, pursued a career in venture capital. He has always been interested in challenging cases, so he still likes to offer his two cents when it comes to critical decision-making.

"So what are you going to do for this guy?" he baits me, between bites, his smile glistening.

"I don't know. I haven't seen him yet," I answer, stating the obvious, not interested in debating the pros and cons of surgical intervention as we have done on so many other occasions with each other and with colleagues, to the point where it feels that we are following a script.

"Well, just call me if you're going to the OR. I'll leave my cell on." He gives up and pulls a second slice out of the box while wondering out loud what he should do to amuse himself while I'm busy. He'll probably go to a café and read a couple newspapers, further broadening his understanding of the world we live in.

We arrive at the hospital and my husband assumes the driver's seat. I leave the car, enter the hospital through the automatic glass doors that seal shut behind me, walk down the hallway to the ICU, and prepare to immerse myself in a very small, intense, and isolated world.

I have three priorities: evaluate the scan, evaluate the patient, and then step back for a gestalt, big-picture view of everything. The patient's scan is a textbook example of a major stroke that has swollen aggressively. An "ischemic" stroke—due to blockage of blood flow to a portion of brain—causes part of the brain to die. (Just as a heart attack causes a portion of heart muscle to die.) Dead brain tissue swells. If a large portion of the brain has died, such as in a complete middle cerebral artery stroke, the swelling can be quite impressive, to the point where it causes the brain to shift within the skull, threatening the viability of the remaining normal brain and brain stem.

Where I trained as a resident, a "strokectomy" was advocated in certain life-threatening situations. In a strokectomy, a portion of dead brain tissue is surgically removed in order to leave more room for the remaining, unaffected brain. The concept is somewhat controversial and the practice is not widespread, but it truly can be a life-saving procedure. The question, of course, is whether or not the life in question *should* be saved, given concerns of quality over longevity, and that's where the decision can get tricky.

I did a brief neurological exam on the patient while the family waited outside the room. He was clearly in dire straits. I ordered a stat dose of mannitol to be given through the IV to buy us some time. This would temporarily lower the intracranial pressure by dehydrating his brain a bit. The effects don't last long but it's perfect for such a situation. I would have a few minutes to talk to the family, the neurologist, and the internist in order to get the critical big-picture view, as this was my first time meeting the patient. What was he like? What would he want? What do we want?

A living will doesn't always give clear direction. I remember seeing one in a patient's chart that included a few lines at the end that the patient had added in his own words. It said, in large, almost childlike, handwriting: "I do not want no machine hooked to me. Soley to keep me alive." Unfortunately, I couldn't extract any further meaning from these words apart from what was already in the legal text, but the handwritten words were far more endearing.

Conversations like this require as much listening as talking, if not more listening than talking, and this is when the neurosurgeon is neither scientist nor mechanic. After holding court with all parties, the decision was clear: no surgery. The patient was elderly, and fragile in so many ways. His neurological deficits were significant, and his outlook was bleak, even with a technically successful emergency operation. Everyone agreed to no "heroic" measures (an odd term in a situation like this, as sucking out dead brain tissue in a frail elderly man seems more pathetic than heroic).

In the instant the decision was made, the intensity that had stricken that small world vanished. The nurses moved more slowly.

The looks of panic, fear, and confusion on the family members' faces were replaced by a simple sadness. The neurologist and internist left the ICU to attend to other concerns. I lingered for a few moments in an attempt to soften the blow, as if I could somehow soften the blow.

In thinking like a neurosurgeon, not everything comes down to a mechanistic evaluation of the intracranial contents. You do have to know about everything that can go wrong, and then about everything you can do to fix it, but then you also have to know when to do nothing. Certain decisions come down to a judgment call based on the gut, and that's when both the scientist and the mechanic step aside.

I returned to the nurses' station and dialed my husband's cell phone. He answered and I heard music in the background. I could picture him in the café, sipping a latte and flipping through *The New York Times* "Week in Review" section.

"Can you come pick me up?" I asked him.

"Sure. No surgery? Let's enjoy the afternoon then," he said, matter-of-fact.

And I did enjoy the afternoon, strangely enough. Because, in thinking like a neurosurgeon, you also have to know how to make a decision in the face of tragedy and then just move on.

TWO

Small World

Sometimes I wonder why I chose such a strange career. I don't always have a satisfying answer for myself, but I can accept this lack of clarity from time to time. When other people pose the same question, though, I feel obligated to have a clear answer, so I have developed a respectable dinner party response that takes only a few seconds, something about neurosurgery being the best way to combine my interest in the brain with an interest in doing things with my hands. This answer, though, is admittedly dull and perfunctory, and probably disappointing.

Luckily, most people don't linger over my lackluster response because the "why" question is typically followed by a "what" and a series of "hows": so *what*, exactly, do you do, and *how* do you do this or that? I suspect that other professionals, like teachers, lawyers, and even other doctors, don't get these "whats" and "hows" quite so often, and I think I may know why. The bottom line is, there aren't many of us out there. Not only is the job unusual, it's also extremely uncommon.

I love watching the Parade of Nations during the opening ceremonies of the Olympics. As the athletes stream into the stadium, I read off the name of each country and remark on how many athletes follow along behind each sign. I have a special affection for the smaller nations with smaller representation, like Malta and East Timor. I root for those athletes. In the world of medicine, neurosurgery is one of the smaller countries with fewer athletes trailing behind its sign. Internal medicine, pediatrics, obstetrics, and gynecology, to name just a few—those are all much larger. There's no need to feel sorry for us, though. We're well trained. As a nation, we don't struggle that much. Neurosurgery is similar to one of the Scandinavian countries—small but elite and with an impressive gross domestic product relative to its size. In fact, at large academic medical centers, our economy often helps support the more populous but less economically sound Sudans of medicine. You might consider rooting for one of them.

There are about 4,500 neurosurgeons in the United States. As a visual person, I picture that number as, roughly, all the kids in my high school times two. This means that there is, on average, one neurosurgeon for about every 66,000 people or so in this country. (Compare that to the African continent where the ratio is one neurosurgeon per 1.4 million people. If you exclude Northern Africa and South Africa, the ratio is closer to one neurosurgeon per 6 million people! One piece of advice: avoid a head injury while on safari.) As for the boy-girl ratio, it's still heavily weighted toward the boys. Only about 5 percent of neurosurgeons in the United States are women, but that is changing, slowly, as more presumably intelligent women are willing to subject themselves to a career of demanding and frustrating lifestyle inconveniences.

If an American has never traveled to Scandinavia or is not well acquainted with world geography, he may have difficulty in distinguishing between Norway, Sweden, and Finland on a map. Similarly, unless you have personally dealt with a neurosurgeon before, you may not be able to distinguish between neurosurgery, neurology, and maybe even psychiatry. This confusion stems from the common focus on the brain. The specialties, though, are quite different and each has a strong national pride. Neurosurgeons tend to wince at the mistaken identity. They joke that one good MRI scanner can replace a hundred neurologists. The snobbery here revolves around the assumption that a surgeon's skills cannot be replaced by a machine (at least not yet). I can't comment on whether neurologists are subject to the reverse mistaken identity, but I can attest to the fact that most would consider themselves brighter than surgeons. They are also more likely to have a beard or wear a bow tie, by the way.

Because our jobs are unusual, I need to clarify what we do, who we are, and who we aren't. The "surgery" part is a clue. Neurosurgeons operate. Neurologists and psychiatrists do not. For that reason, neurosurgeons focus on disorders amenable to surgery, which constitute a relatively small subset of all brain-related disorders. That is one reason for our small numbers. Brain tumors can pose a serious threat to individuals, but they are not public health menaces.

So, for instance, Alzheimer's disease seems to be on everyone's mind. Neurosurgeons don't treat Alzheimer's disease because no one has ever designed an operation for it. Consider schizophrenia. It affects 1 percent of the population, but, again: no operation for schizophrenia, so no role for neurosurgery. If you ask me about schizophrenia, I can tell you what I learned in medical school a decade ago or, better yet, what I read most recently in *The New York Times*. I will certainly be conversant, but not an expert. What about these disorders: autism, attention deficit, and Tourette's? Again, these are not within our surgical scope, but I do find them to be a fascinating read.

Neurosurgeons are true experts in brain trauma, brain tumors, aneurysms, congenital brain anomalies, hydrocephalus ("water on the brain"), and brain hemorrhages, among other things. All of these problems have at least the option for a surgical solution. The borders between us and our nonsurgical colleagues, though, are porous in spots. Some diseases can be treated by both a neurologist and a neurosurgeon, like Parkinson's disease or certain forms of epilepsy. A neurologist may refer such a patient to a neurosurgeon when it becomes clear that medication alone is ineffective, or if medication side effects are too much of a problem. Surgery is usually a last resort.

Classically, neurosurgeons complain that neurologists wait too long and try too many medications before sending patients over. Neurologists complain that neurosurgeons are too aggressive in recommending surgery. Such grievances keep things lively.

Leaving the brain aside for a moment, you may not realize this: the majority of neurosurgeons spend the majority of their time operating on spines rather than brains. This may be a minor disappointment, because the spine tends not to inspire the same degree of intrigue and appreciation as the brain. The predominance of spine surgery is simply a reflection of what's out there in the population (and, some cynics would argue, what tends to reimburse well). You probably know someone with a "bad back" or a "bad neck" and may even be a victim yourself. Arthritis of the spine (or degenerative spine disease, as we like to call it), unlike brain tumors, *is* a public health menace.

Because of this broader focus on both the brain *and* spine, we don't refer to ourselves as "brain surgeons." The term is too narrow. Also, it sounds silly for some reason, so it's hard for us to say it with a straight face (although I have heard of neurosurgery residents using the term as a pickup line in bars). Likewise, rocket scientists don't actually call themselves rocket scientists, but I'm not clear on what they do call themselves.

One man is historically credited with having inspired the enduring "brain surgeon" image in the public eye. Harvey Cushing is the widely acknowledged father of neurosurgery. To give you a sense of the field's youth, Cushing was born in 1869 and died in 1939. He was really the first person to perform brain surgery in a thoughtful and systematic fashion despite the overwhelming surgical mortality rate at the time. Most surgeons were intimidated by the thought of operating inside the skull. Cushing, though, was a pioneer and a bit of a renegade. The techniques he developed and the rigor he brought to the discipline made brain surgery a reasonable endeavor for others to pursue. Neurosurgery is its own specialty largely because of Cushing.

What made Cushing even more remarkable, though, was that he was more than just a pioneering neurosurgeon. Although his clinical skills made him famous among physicians, it was his literary skill that was instrumental in his recognition by the larger public. He won a Pulitzer Prize for a biography that he wrote about another famous physician, William Osler. This and his later writings made him a well-known entity in the literary world and gave him coverage in popular publications like *The New York Times*, *Newsweek*, and *Time*.[1]

After Cushing's death, his own biography was written by one of his protégés, John Fulton. I have an original copy of the book, published in 1946. I bought it at a used bookstore in Cleveland—my hometown—the city that Cushing grew up in.[2] On the inside cover, written in pencil, is the following: "Abram Garfield, from Hope & Ted, Xmas '46." On a whim, I looked up Abram Garfield in the index when I bought the book, and noted four separate pages listed after his name. The former owner of my book was Cushing's friend.

Cushing's biography is over seven hundred pages long. Although I have never read through the entire tome, I did make sure to read the last couple pages which, I discovered, are of particular significance to neurosurgeons. In true surgical style, Fulton describes Cushing's autopsy in a fairly matter-of-fact clinical fashion:

"Drs. Milton Winternitz and Harry Zimmerman reported that

the brain showed no sign of atrophy but the arteries were here and there sclerosed; and in line with the superstition that physicians sometimes fall victim to the diseases in which they specialize, a small colloid cyst, one centimeter in diameter, was found in the third ventricle."[3]

A colloid cyst! Harvey Cushing, father of neurosurgery, harbored a colloid cyst deep within his own brain. (I don't think even many neurosurgeons know this.) A colloid cyst is a quirky little entity. It's completely benign in the sense that it's not a tumor or cancer, but it is associated with the possible risk of sudden death. It sits at a critical crossroads of cerebrospinal fluid circulation in the brain such that if the cyst grows large enough, it can block this flow and cause a dramatic and rapid increase in the pressure inside the head. Luckily, colloid cysts are pretty rare. In Cushing's case, the cyst was an incidental finding. He probably died of a garden-variety heart attack. He was a smoker.

Colloid cysts are fun to remove. A gelatinous goo often oozes out from the center after you pop through the cyst wall. I say this with the realization that most people wouldn't want to think of their neurosurgeon as having "fun" (or marveling over gelatinous goo). But you have to have some fun with your job, or why do it? As a junior resident, I was privy to an unusual conversation between one of the academic neurosurgeons and a woman with a newly discovered colloid cyst. The neurosurgeon explained that he was going to have to refer her to a colleague in the department who was more of a specialist in the endoscopic technique recently advocated to remove such cysts. He then added: "But don't get me wrong. . . . I'd love to take that sucker out!" The woman appeared dumbfounded as the neurosurgeon patted her on the back and walked out, leaving us in the room together alone. An awkward silence lingered. Some things are better left unsaid.

Cushing was buried at Lake View Cemetery, a historic cemetery in Cleveland that is also the final resting place of J. D. Rockefeller and former president James Garfield. When my husband and I were still medical students, we wandered around the cemetery in search of his grave. We found it only after consulting a cemetery map, as it wasn't one of the most impressive tombstones. As eager medical students, we had become Cushing groupies, several decades too late.

I can assure you that, as a kid, extracting a nail from someone's head or wandering around a cemetery in search of a neurosurgeon's tombstone were not activities I would have envisioned for my future. I was, though, privy to the world of surgery well before I made a single career move, and this must have affected the wiring of my brain in ways undetectable to my child mind at the time.

THREE

Influences

I grew up with surgical stories. My father is a surgeon—a general surgeon—and I understood from an early age that there was some degree of gore inherent in his job. This was a great source of pride for me as a kid. Most of my friends had fathers who sat at desks all day. What kinds of stories were there in that? I imagined my friends, helpless, subjected to lifeless desk-based tales at the dinner table. Lucky for them, though, I was happy to spread my wealth of second-hand tales from the trenches.

The account of the poor kid with the hair bezoar became legendary. This was a child my dad took care of when he was a resident. The word "bezoar" is probably one of the ugliest words in the English language, and rightfully so. It refers to a wad of any given material that gets stuck in the stomach, often requiring surgical extraction. Typically, this wad collects slowly over time. The word "bezoar" is not used in everyday speech, which makes it all the more intriguing when used on the rare appropriate occasion, as with this story.

The subject of this nonfiction was a boy whose grandfather, I believe, was a barber. The boy hung around the barbershop regularly, and supervision was apparently lax. Between customers, he had a peculiar habit of crawling around the perimeter of the room, picking up scraps of hair off the floor, and swallowing them. Over time (was it months, or years? I can't remember now), this child developed a massive hair bezoar in his stomach, right under the unsuspecting noses of his family and friends. When things finally came to a head, with the boy having no room left for conventional nourishment, he required major surgery to extract the mass. When the family learned of the contents, they pieced together the chain of events with disturbing clarity.

This bezoar had taken on the exact shape and size of the boy's enlarged stomach and my father had the foresight to capture the image of the extracted hairball on a thirty-five-millimeter slide. Surgeons like to show each other pictures when they give lectures, partly for teaching purposes and partly as a thinly veiled version of the classic "here's what I did on summer vacation" show-and-tell. In surgery, more often than we'd like to admit, anecdotes rule, and all the more powerfully when accompanied by high-quality Kodachromes. Surgeons know that in giving a lecture, it is also important to include the results of scientific, controlled, multicenter studies. The accompanying charts and graphs, though, tend to be either confusing or boring, so color pictures of things removed from patients' bodies are snuck in whenever even remotely justifiable.

As I grew older, it dawned on me that the story of Bezoar Boy was more pitiful than captivating. The case falls into a psychiatric disorder known as "pica," or the involuntary urge to eat nonfood items. I was recently reminded of this child when I read a short news clip about an elderly man in France with a penchant for eating coins.

Over several years, his stomach became so stretched out and weighted down with coins that it descended into his pelvis. During surgery, upon opening the precariously thin-walled, flimsy organ, the surgeons noted disintegrating francs at the bottom and—logically—euros layered at the top. This detail takes the cake and rivals anything I could possibly tell you about the aforementioned hair bezoar. The case is equally pitiful, though, and I feel a bit guilty in exploiting the old man as a supporting anecdote.

My father, in telling his clinical accounts, was always respectful. The patients remained anonymous. He told the stories with a sense of wonder, not mockery. As a kid, I felt privileged to have such insight into the extremes of human behavior. I sometimes thought about his patient who swallowed a pen in jail, just so that he could leave his cell for the hospital, only to do it again upon his return. Talk about pitiful. My father reflected on the inmate's state of mind, shaking his head in disbelief, and this story became my definition of human desperation.

Looking back, there was another hint of my father's dignity as a human being, more clear to me now as an adult: he forbade me from going to the so-called freak shows at the summer county fairs in Ohio. We lived in Shaker Heights, a nice suburb of Cleveland, and we visited these rural fairs to gain an appreciation of a different scenery and culture, and for the greasy funnel cakes with powdered sugar. Although there were plenty of other things to see—livestock, choice vegetable specimens, Amish buggies in the parking lot—I always begged my parents to let me go through the Hall of Human Wonders. At the time, I didn't understand what was wrong with paying money to see the world's fattest woman, or the wolf man with excessive facial hair, or the rubber man (who I now know probably had a serious, and potentially life-threatening, connective tissue disorder). Although my father was willing to divulge the true story of Bezoar Boy in the privacy of our dining room, he drew the line at paying to gawk at unfortunate oddities.

My younger sister Ingrid didn't share my interest in gawking. I attributed this to her relative discomfort with human deformity as a youngster. We frequented a Chinese restaurant at a strip mall where one of the waiters was missing a hand, and she never ate much there. I, on the other hand, reached for seconds while marveling at his ability to balance plates with the aid of his stump.

Although I'm sure my father included plenty of uplifting stories in the dinner table mix—the archetypal ones about medicine and miracles, about patients with hope defying the odds—my child's brain filed away only the more graphic and disturbing ones, in permanent storage.

Apart from my access to gory stories, I took pride in my surgeon-father for another reason: he could handle anything. When it came to the threat of backyard injuries, I perceived my friends to be at greater risk of delayed or inadequate treatment, given their fathers' narrow desk-based skill sets. I wondered if that made them nervous. What could their fathers do, aside from applying a Band-Aid? I knew that my dad, on the other hand, could always perform emergency surgery if he absolutely had to. He remained calm in the face of active bleeding. For any given backyard or playground trauma scenario, he had seen worse, much worse.

Take, for example, the kid who ran with scissors. My father strongly cautioned against this activity, and he only had to tell me once. This simple lesson lodged itself in my mind, indelibly, because it was accompanied by another slide from his personal archives. This single picture was equal to a childhood's worth of warnings. The slide showed the face of a young boy, a patient of his years ago, who actually did run with scissors—and tripped. In the small slide, held up to the light in the kitchen, I could see the gleaming handle of the scissors protruding from the corner of his eye socket. Needless to say, from that point forward, I was reluctant to run with anything sharper than a spoon. However, had I become the victim of a scissors injury, I knew my dad had seen it all before and would know what to do.

Although my father never did have to perform emergency surgery at the kitchen counter, we had some of the necessary tools lying around, just in case. For some reason, we always had a weighty pair of curved shiny scissors in a kitchen drawer, packed in alongside other random utensils. Although I suspected that their original and proper home was an operating room, and that they had somehow made their way into our home, I didn't know that they had a name—curved Metzenbaum scissors—until years later when I became a medical student. These scissors were, at the time, an anonymous annoyance. Although they were fine for low-precision tasks like cutting apart clumps of grapes, they were lousy when it came to wrapping gifts. They created an awkward scalloped edge to the paper that grated on my perfectionist tendencies.

My mother would trim off the ends of my straight hair every few months, and I was charged with the task of hunting around for a good pair of scissors, which often took longer than the haircut itself. Without fail, I would think my search was over when I spotted the shiny handle of a pair in a drawer, only to realize, a moment later, that I had been cruelly misled by the curved ones. In this realm, my otherwise disadvantaged friends had at least a minor advantage over me, as they did not have the rogue operating room instrument contaminating their household items.

I remember goofing around with my siblings, throwing things other than laundry down the second-story laundry chute in the hallway that led to my bedroom. Somehow, I slammed my finger in the door of the chute—hard—and the pain was throbbing and unbearable. I ran downstairs to report the injury to my dad and promptly fainted. I awoke, startled, to the noxious stimulation of smelling salts under my nose. My father kept some of this substance in a black doctor's bag in a library cabinet, expressly for this purpose. I was thankful for the rapid wake-up. I suppose, in retrospect, that I would have regained consciousness on my own just fine, but the smelling salts were a nice touch and further supported my pride in being the daughter of a surgeon. I haven't seen smelling salts since, even in a hospital. Its utility must be under question.

Despite having a surgeon as a father, with all of its advantages, surgery was not exactly the job I had in mind for myself as a kid. Some children know they want to be a doctor, and even specifically a neurosurgeon, as far back as elementary school. They announce their plans to everyone, including strangers in line at the supermarket, and when they grow up to become a doctor, nobody is surprised. I wasn't one of those kids. Maybe I didn't like the thought of missing dinner on a regular basis, working on weekends and holidays, and spending inordinate amounts of time in a standard issue hospital. I was happy to hear the gory stories, but I didn't aspire to have my hand in them.

Although I had my own career uncertainties, there was always an expectation among my extended immigrant family that I should, and would, become a doctor. I was the oldest of four children, and though the younger kids weren't necessarily off the hook, the relatives worked on me first. For my birthday once, an aunt gave me a little doctor figurine—a middle-aged male figure in a white coat. The figurine didn't really speak to me. It might as well have been a firefighter, or a bullfighter. I decided not to display it on my bookshelf alongside my shells, flashy chunks of fool's gold, and other highly prized artifacts.

I didn't know what I wanted to be. I knew what I liked, but my interests didn't add up to any particular profession that I could display on a shelf. I had a natural curiosity for anything foreign. I collected stamps and coins from as many countries as possible. Monetary value was irrelevant. Obscurity, or exotic design, was what I was after. I wrote to pen pals around the world: France, Thailand, the Congo. (I dropped the one from France when it became clear that she was obsessed with our native Sylvester Stallone, and kept asking me to clip and send any pictures I could find. I felt used.)

I penned a letter to the president of the tiny Pacific Ocean island country of Nauru. Our encyclopedia, which I flipped through for entertainment, included a disappointingly small entry for this country, but did mention its admirable per capita income owing to rich phosphate reserves from bird droppings. I had to know more. I never received a response but blamed the post office rather than the president, figuring they didn't know what bin to toss my letter in. I have since learned that Nauru's phosphate reserves have nearly been depleted, leaving large craters in the landscape that dim future prospects for tourism.

In junior high school I decided, on my own, to make a list of every country in the world and each capital city. I memorized the list. I wanted to know at least one fact about every country, including the lesser-known Tuvalus and Bhutans, and felt this was a good place to start. I prided myself on this specialized knowledge, a knowledge that other kids didn't have (or didn't care to have). In retrospect, the exercise was a bit obsessive-compulsive, but I'm still pleased to know that Antananarivo is the capital of Madagascar.

In addition to my obsessive focus on the foreign, spending time outside was a priority for me. I enjoyed climbing trees, walking through the knee-high Cleveland snow, and exploring any semblance of woods in our suburb. In high school, I was traumatized by an article that compared the amount of time adults and children spend outdoors. I wanted to remain more childlike in that respect, and a traditional indoor job would be a declaration of defeat. Along with the rest of my high school class, I took a lengthy personality survey. The goal was to match interests, skills, and personality with potential career choices. A few months after answering the several-page litany of multiple-choice questions, I received my own personal career list. My ideal career choices, determined by mathematical algorithm, were ranked in order of best fit. "Forest ranger" was number one. "Physician" was on the list, too, but somewhere down in the teens. I did recall marking "working outdoors" as a priority, but hadn't considered that the options were limited.

I had no idea how to combine my interest in foreign cultures with the outdoors *and* make a decent living, but I started on the culture focus by majoring in anthropology in college, and signing up for Japanese language classes (requiring several hours of instruction, and several more of practice in the language lab, per week). Just prior to college, I spent part of the summer living with a family in Nara, Japan, as part of an exchange program. My time there was a revelation: I was meant to be born Japanese. I loved the culture, food, traditional architecture, orderliness, Buddhist influence, and the way they wrapped even the most mundane items bought at almost any store.

Partway through my sophomore year, though, I started to fixate on the more practical concerns, like making a good living and maintaining a stable career. I remained an anthropology major but became pre-med, which forced me to quit my demanding Japanese classes just at the point where I had become comfortable writing my own name. It was depressing to think that I had come so far but had to quit early. I would always remember how to ask for directions to the bathroom (using either the formal or informal term for bathroom), but I might not understand the response, which minimized the value of knowing how to ask the question in the first place.

In part, I have to thank (or blame) my then-boyfriend and now-husband, Andrew, for this change in focus. He was one of those kids, at a very early age, who announced to strangers in the supermarket that he was going to become a neurosurgeon. He convinced me that pre-med classes at Cornell would be no problem. After all, he was certainly breezing right through them himself. But my Mr. Smarty Pants was off the mark a bit. Organic chemistry was a painful exercise in the short-term memorization of shapes, letters, and numbers and the complex system of all those things merging with one another. The class was held in a massive two-tiered lecture hall with hard-core savants seated all around me. Physics was only a little less agonizing. Andrew did what we still refer to as the wild-limbed "physics dance" when I told him that I got an A— in the class. Unfortunately, I didn't earn an "organic chemistry dance."

I returned to my hometown of Cleveland for medical school and lived with my family for the first year. My youngest sister, Elizabeth, (fourteen years younger and in elementary school) was at a very curious age and soaking up knowledge like a sponge. After being away for four years of college, I was eager to get back to telling her bedtime stories. Very quickly, though, I exhausted my sketchy repertoire of the classics and decided to transition to a less traditional theme, but one that was more current in my mind: parasitology. I assumed that cardiac physiology and neuroanatomy were just beyond her, but my new expertise in human parasites from around the world proved to be just her speed.

She accused me of telling a tall tale when she heard the one about *Dracunculus medinensis,* the parasite that causes a disease called dracunculiasis in Africa. I explained that the female worm is long and thin, up to three feet, and usually lives just underneath the skin of the human host's legs or feet. When she is ready to release her offspring, a skin blister develops, the blister turns into an ulcer, and the end of the worm is exposed to the outside world. In order to remove the long, thin parasite, locals afflicted with this worm wind the exposed end around a stick, and turn the stick, slowly, so as not to break the worm, until the whole thing is released. Despite her suspicions, she requested a retelling of this one at a subsequent tuck-in. Her unflagging curiosity was an inspiration.

Perhaps my proudest moment, though, as an older sister featured an even lowlier parasite, the pinworm. A couple years had gone by, and I assumed that the parasite parables had faded from her memory. My family was invited to a friend's home for dinner one summer night, and Elizabeth left the dining room table, mid-meal, to use the bathroom. Upon returning, she stood behind my chair and whispered into my ear: "I think I have pinworm." I left the table and pulled her aside in the hallway. "How do you know?"

She explained that she had been itchy lately and was certain that she saw one of the tiny worms in the toilet, just then. Plus, she was attending summer camp along with plenty of other grimy kids with questionable hygiene and eating habits—the perfect setup, as she had recalled. She even remembered the simple confirmatory "Scotch tape test" that picks up the tiny eggs, commonly deposited around a certain orifice, when pressed against the skin. I told her that we would leave that to her pediatrician.

My mom had never heard of pinworm. However, based on the steadfast conviction of her youngest and the support of her eldest, she was willing to suspend her suspicions, be a good sport, and take a trip to the pediatrician's office anyway. The diagnosis was confirmed. Elizabeth then rid herself of this annoying, but fairly harmless, infestation upon the ceremonial swallowing of a single antiparasite pill. I beamed with pride, as any big sister would under similar circumstances.

Although I briefly—for about two days—considered going into parasitology as a career, my thoughts changed. The creatures were intriguing but I couldn't envision building an entire career around them. The first couple years of medical school are like a buffet, where you get a little taste of everything as you go around the table. But eventually, you are presented with a menu and can pick only one dish. That can be difficult for students who are generally and widely curious. In choosing one area, you are excluding the others. As a way of hedging your bets, you can choose a general field, like internal medicine. I didn't consider that option. I suspected that the specialists got the most flavorful dishes, leaving the internist with the bland staples, like rice. I valued the notion of serving the public, but I couldn't get excited about treating high blood pressure day in and day out.

In college I deliberated over entering medical school, but once in medical school, I deliberated very little over neurosurgery. The brain was definitely more interesting than the kidney (or the heart, or the bones, or the skin . . .). The kidney balances electrolytes and produces urine. The brain harbors personality and produces thought. If push comes to shove you can use someone else's kidney; there's nothing unique about your own. Your brain, on the other hand, is who you are. So, when presented with the menu of options, I knew that something brain-related was required to feed my curiosity over the span of a career.

The stories of neurologist Oliver Sacks gently tapped me over the edge in the direction that I was already going. His famous book, *The Man Who Mistook His Wife for a Hat*, featured a quirky guy (Dr. Sacks) noticing quirky things (unusual neurological symptoms in his patients). He was passionate, inquisitive, and thoughtful. Stories like these, I figured, could only be written about the brain. To this day, I have not come across equally intriguing tales concerning the pancreas.

I asked other neurologists what they thought of Oliver Sacks. Some felt he was a good storyteller but a run-of-the-mill neurologist. I found these comments cruel and unfair, first because these neurologists didn't really know him and, second, because they were probably just jealous. They had certainly never written anything about the brain compelling enough for medical students (and the lay public) to devour during their free time. To my disappointment, a few neurologist-cynics even seemed to have lost their sense of wonder about the brain, which can happen, understandably, to a volume-dependent service provider whose days are packed with fifteen-minute visits, especially once they've been sued over one of those fifteen-minute visits.

Despite the influence of Sacks, though, I decided against neurology. There was no real manual component to the job, and the mainstay of intervention consisted of prescribing medication, which I worried might not sustain my interest. So I looked into neurosurgery instead. I remain a big fan of Oliver Sacks, though. A highlight of my neurosurgery residency was driving him back to the airport after he gave a lecture in town (our chairman knew him from decades earlier and was able to arrange for this chauffeur role, to my excitement). Andrew and I had a chance to sit down with him, discuss a few philosophical quandaries, and buy him a drink. (In his absent-mindedness about mundane matters, he forgot to bring his wallet on the trip.) True to the literary persona that I had come to know, he was quick to notice and comment on anything peculiar, like the word "standee" on the sign posted over the walking sidewalk in the airport ("Walkers to the left. Standees to the right.").

Returning to my days as a medical student with a menu in hand, I was faced with a final decision: What specialty? I considered my influences, my curiosity about the brain, and my newly emerging inclination toward surgery. Then, with little thought given to the lifestyle consequences of the most critical career move in my life, I decided to become a neurosurgeon.

FOUR

Acceptance

If you hope to become a neurosurgeon, you have to prove your passion for science to the gatekeepers in the academes of neurosurgery. Even if the manual work is what really excites you, it would be unwise to say, "I just want to open heads" during your interviews. Surgeons are fond of explaining that they could teach a monkey to operate. What they mean is that operating is only part of what a surgeon does and thinks about, and it's not the hardest part. I remember my father recounting this classic monkey line when I was too young to appreciate the role of sarcasm in conversation. The literal image was unsettling.

At the most competitive residency programs, a mind for science is particularly high on the list of essentials. These programs aim to continue their tradition of prolific research publication, and residents play a key role in cranking out the papers. Program chairmen also want their graduates to remain in the academic system after residency. The thinking here is that the most serious scientist-surgeons are less likely to be lured into lucrative private practices. Regardless, smart graduates often want or need to take advantage of their most lucrative free-market options anyway, especially the financially strapped ones with growing offspring. Medicine is a business, after all, as well as a profession.

So, recognizing the desirability of the scientific mind, savvy medical students do all they can to bolster the research sections of their résumés—the academic equivalent of a peacock's tail—often to the detriment of their already tenuous social lives. Fear, in addition to savvy, is a factor here, too, as there are always more applicants than spots, similar to musical chairs. Neurosurgeons were those kids who always managed to grab hold of a chair.

In anticipation of becoming a neurosurgeon, I, too, spent a few afternoons a week in a laboratory as a medical student so that my résumé would excite the decision makers on the interview circuit (or, at least, would avoid inciting laughter). It was worth it. If you were to examine the research section of my résumé, you would notice the following title of one of my projects: "Use of amiloride to minimize reperfusion injury in gerbil model of ischemia." You must admit that the words "amiloride" and "ischemia" are intimidating. Perhaps "reperfusion" gave you pause, too, but you may have seen that word in other contexts. Although I did do my best to come up with an impressive line on my CV, I was trumped by my competitive colleagues who had the foresight to choose a research project in genetics, thus qualifying them to include formidable nonwords such as "p53" in the titles of their projects. (This has nothing to do with page numbers.) There's no way to prove it, but they may have had a slight edge over me.

Now that more than ten years have passed since I spent sunny afternoons in a dark lab during medical school, I have to disclose which word in my research project title had the most enduring impact on me: it wasn't "amiloride" or "ischemia," but, rather, "gerbil." My research project involved trying my best to cause strokes in these subjects, animals that other people call pets, not subjects. The long-term goal was to prove that the drug could minimize stroke damage in humans whose brains were at risk during certain operations. Unfortunately, the drug had to be given just *before* the stroke, so wider and more practical outside-world applications were hard to envision.

By the way, I do understand the emotions of those who would oppose such use of rodents in medical research, but I am swayed far more by the emotions of a family gathered around the bedside of a stroke victim, wondering why nothing more could be done.

In order to cause these strokes, I was taught to carefully and completely cinch off the two most significant arteries that feed the brain, the carotid arteries. Even then, despite such extreme measures, the strokes ended up being pretty small. If only humans were so resilient. It always amazed me how hearty a gerbil's brain was in comparison to that of Homo sapiens. And for what? Evolution really gypped us.

Creating gerbil strokes requires a mini-operation performed by mini-instruments on a mini-subject. All told, this was the most satisfying aspect of the project for me. It proved that I had at least the basic manual skills—the monkey part—to be a surgeon. Previous tasks in my life that required coordination, such as mastering the use of chopsticks or playing the piano, were not as clearly translatable. My confidence was bolstered. Never mind that the young friendly lab tech who taught me the technique could do the job just as well, and maybe even faster.

The most disturbing part of the project was not creating the strokes, believe it or not. I could justify that in the name of science and future human welfare. What did get to me, though, was the method of ending the gerbils' lives. The scientific protocol required the freezing of their brains, literally and figuratively, at a precise moment just a few minutes into the stroke process. Here's how it's done:

you take a limp anesthetized gerbil by its tail and lower the entire body, head first, into a large silver cylinder filled with liquid nitrogen. It comes out frozen stiff.

Regarding the wonders of liquid nitrogen, I heard from the more seasoned lab researchers (the so-called lab rats) that if you submerge a rose into liquid nitrogen and then drop it on the floor, it shatters into numerous pieces. (Is that what these guys did after hours? What else were they submerging?) With that image in mind, I was especially firm in holding on to the gerbils' tails upon lifting them out of the cylinder.

After the freezing, I rolled each subject into tin foil like a burrito, labeled them by subject number (they didn't get names), and placed them in a freezer alongside other people's subjects. During my next afternoon in the lab, I retrieved my gerbils for the following step: isolating the brains. This required chipping away at their thin skulls with a scalpel, taking care not to violate the underlying cortical brain surface, an inelegant and tedious task similar to whittling a stick, but more precise. I had to hold them with an oversized oven mitt to protect my warm hands from their frozen bodies and vice versa.

It could be that my distaste for the project was rooted in family history. One of my younger sisters always had gerbils as pets. Although I didn't feel much affection for them myself, I respected the fact that she did. I felt bad when my father had to put one of them to sleep and my sister cried. (He brought home some sort of gas from the hospital. The gas plus a brown paper lunch bag was all that was required to perform this simple act of mercy, which reached a level of urgency when the sickly animal started to gnaw at its own hands.) So much emotional trauma surrounding one single small-brained rodent, and there I was, years later, dunking one after another into liquid nitrogen so that I could become a neurosurgeon.

As a physician, I understand and respect the critical role of selective animal experimentation in advancing science and medicine. Rodent martyrs may well contribute to my own future health and longevity. I learned from my gerbil experiment, though, that I would prefer to leave such critical work to the "lab rats." All of my projects from that point on were based on human data. Most relied heavily on the civilized review of room-temperature medical records.

Even the seemingly guilt-free projects that I chose, though, the ones that required no sacrifice of gerbils or other living beings, could be a little uncomfortable at times. In an effort to help me get into a good program, a mentor of mine suggested I do a study that would be quick and straightforward: look at a small group of patients with an uncommon type of tumor and see how they did after neurosurgical intervention. The sample size would be small. All I had to do was get the charts, organize the data about the patients, their tumors, and their treatment, and look at outcomes. Outcome meant length of survival, which seemed simple enough. I could have a paper done in no time.

The catch here was that many patients were from out of town and had much of their follow-up elsewhere. They had traveled to the big university for specialized treatment but it was too far for them to return again and again. Some of the charts clearly stated the date of death (a short note usually written in the neat secretary's handwriting rather than the more cryptic surgeon's hand, based on information from a phone call or newspaper obituary), but others left me hanging.

One chart documented Mrs. So-and-so's six-month follow-up visit. She was stable. I could almost picture her, in the surgeon's office, grateful for her stability. That was two years earlier, though. The chart had remained filed, stagnant, until that day when I happened to reach for it. Mrs. So-and-so was about to play a small role in this incremental step toward achieving my career goals, and I was worried. I didn't even know Mrs. So-and-so, but I assumed the worst and felt sorry for her husband whose name I had spotted on the face sheet in her chart.

I went back to my mentor and asked what to do with all the patients who fell off the radar screen. "Just call them up!" was the reply; an obvious answer that made my question seem ridiculous. Easy for him to say. He wasn't the one who was going to have to ask for Mrs. So-and-so, only to endure a long pause from the other end of the line, from her widower. What was worse, not only did I have to find out whether or not a patient was still alive, I had to get the date and cause of death.

I did what I had to do. Luckily, most of the families were not only helpful but gracious. Still, I felt just a little awkward. In a small way, I was relying on these families to help me get ahead. I kept reminding myself that they were contributing valuable information in the name of science, not just in the name of my CV.

I hoped that, in the end, all of my science projects would pay off, landing me a spot at a top-notch residency program. The final selection process, though, was a black box for me as a fourth-year medical student. Not only was it a black box, it was also frighteningly out of my control, especially once all the variables that I could influence or tweak—test scores, research, papers in press, glowing letters from mentors, application essay—had already been influenced or tweaked. Interviewing was the final step. The decision process, after that, remained a mystery.

At least everyone else was in the same boat. Midway through the fourth and final year of medical school, a pack of medical students hoping to become neurosurgery residents spend a wad of money (that they don't really have) flying around the country to interview. The next couple months after that are spent worrying. You tend to see the same students over and over again, and as you watch a certain guy turn on the charm with all the decision makers, you tend to wonder: Is this the guy who will be taking my spot here? Most programs accept only one or two new recruits per year, so the paranoia is actually rational, not a precursor to schizophrenia. You know that everyone else's CV must be as good as or better than yours (they got the interview, too), and so charm can be a critical distinction.

As a woman trying to enter a largely male-dominated specialty, what was I supposed to wear to my interviews? I didn't have many mentors to turn to in this deceivingly trivial dilemma. I had interacted with only a couple female neurosurgeons up to that point. One was exceptionally smart but a bit frumpy, and the other, although also smart, wore higher heels and tighter skirts than I could have managed. And one female neurosurgery resident I knew had a brain tattoo on her hip, before tattoos were commonplace, and I wondered if she used it as proof of her dedication during interviews. (She eventually left neurosurgery to become a radiologist.) Regardless, I wasn't willing to go that far.

I took the conservative approach with a dark pantsuit, always a safe choice. I still struggle at times with matters of style. I would like to branch out, but I feel a bit constrained. When a sales clerk suggests a great looking but low-cut blouse, I am tempted to explain that patients do not appreciate a hint of cleavage in their surgeon. It does nothing to inspire additional confidence.

Interviewing lore is passed around from program to program and from year to year. There was the chairman who claimed to offer a spot to anyone who could beat him in chess. There was the student who split the back seam of his pants in the bathroom and, in his contortions within the stall to try to remedy the problem, managed to dunk the end of his tie into the toilet. There was the guy who completely mangled his interviewer's name, and so on.

On the interview circuit, we warned each other about which neurosurgeons tended to ask tough questions and which might actually put you on the spot with an anatomy quiz. I was most impressed by the audacity of one student who was asked to draw a detailed cross-section of the spinal cord (a complex structure) and label all the parts. He drew a simple circle within a circle, pointed to the inner circle with his pen, and stated: "Filum terminale." Although the filum terminale *is* technically a part of the spinal cord, it's really just the simple, spindly, nonfunctional tail end of it. I don't recall if his cockiness worked to his advantage or disadvantage, but I could see it having gone either way.

Despite the lore, most of the questions posed to me were straightforward: Why do you want to be a neurosurgeon? Where do you see yourself in ten years? How do you know you can handle the stress? And (if they didn't bother to read my application), what kind of research did you do? One question, though, did stick out as more amusing. It seemed specifically tailored for me: How do you know you can handle all the big drills? I smiled and assured the older, male interviewer that I could handle the big drills. Short of performing a demonstration in his office, that was about all I could do to address the question. And, over the next several years, I had the chance to use big drills on several of his own patients, settling any concerns he may have had.

In general, I don't bother getting worked up over minor things that could be construed as sexist. Most people (myself included) don't enjoy working with colleagues who are alarmist, easily outraged, or overly sensitive. I prefer to prove my abilities, naturally, over time. Luckily, in this modern era, I've never found the need to storm out of a room, call anyone a chauvinist, or report any transgressions to the authorities. The way for women in surgery has already been paved to a great degree, and I'm grateful for all the women who must have had it harder—much harder—than I did.

(In some ways, my experience in residency was actually easier than most because my husband is also a surgeon, so he understood the erratic schedule and the commitment required. We often had dinners together in the cafeteria, sometimes took call on the same nights, and even operated together, on occasion.)

Aside from being a woman, something else about my application apparently stood out: I did not use up the entire space allotted for my personal statement. Many neurosurgery applicants, obsessive by nature, tend to fill the entire page in the smallest possible font. I filled the majority of it, but stopped when my point had been made, which wasn't at the very bottom. I left a wide margin. One interviewer told me I had one of the best statements he had read. I'm convinced that was because it was shorter than most. That's one piece of advice I can give with confidence: on an application, more words are not necessarily better. In fact, when an interviewer is faced with the task of reading through dozens of applications in one sitting, brevity coupled with clarity is much appreciated.

The black box of choosing candidates for neurosurgery programs was finally revealed to me when I was given the opportunity to participate in this hallowed selection process as a chief resident, coming around full circle after seven years of training. Our program was more progressive than most in that the faculty actually considered our opinions in their decision-making.

What struck me was our conversation regarding an applicant with a preternaturally strong science background. His résumé fanned out as the most impressive peacock tail of the group. He had done everything right: joined the best lab and first-authored a paper slated for publication in a well-regarded journal. He had big research plans for the future. The most senior and academic-minded faculty members marveled at his résumé and put him at the top of the list. Given the early and hearty endorsement by our leaders, a few other faculty members piped up in support as well (even if they didn't fully understand his research projects). It seemed like a done deal. A cool, seasoned voice from the back, though, dissented: "Listen. Is this a guy you really want in your foxhole?"

The foxhole analogy was an obvious nod to the long-standing comparison between surgical training and military duty. The neurosurgeon I quoted, now retired, was widely renowned for his winning personality, surgical skill, and calm demeanor. The residents revered and respected him. They thought he was cool. So did his patients. He did not have the lengthiest list of publications, but that never seemed to matter. His résumé was not his focus.

With this interjection of dissent, I and the two other chief residents felt licensed to speak up. We were impressed, but not blinded, by the applicant's résumé. Although he was certainly smart, we thought he was kind of a nerd, to put it bluntly, and a bit too arrogant for comfort—a worrisome combination, especially in the trenches. We knew he would be a star in the lab, but how would he be in the OR, the ER, and joking around on rounds?

The faculty heeded our concerns and changed their minds. They sided with us and the foxhole commentator. The science star was moved lower down on our rank list and ultimately ended up at a different competitive program. Was he our loss? It depends on how you balance the priorities: scientist versus teammate.

Overall, I would say that our foxhole mentor was more typical of the average neurosurgeon doing most of the neurosurgery out there, outside of academia: generally smart, skilled, and hardworking, well aware of the standards of care in treating patients, but likely to gloss over half of what is published in the neurosurgery journals. I have to admit that I, too, often give little notice to the same half.

Our journals are roughly divided into two major sections: clinical papers, based on real patients, and laboratory research papers, which are often laced with super-specialized jargon. The latter is the less popular section. I had an interesting conversation with the editor of one of these esteemed neurosurgery journals several years ago. He admitted that even he didn't understand many of the laboratory research papers, but they certainly looked impressive in the journal, and that was important.

This is not to say that the laboratory science is unimportant, of course. Some of the work will certainly save, lengthen, or improve lives in the future. My point is simply that such work is not widely read by the average neurosurgeon whose sole focus is taking care of patients, right now. If a basic science concept comes to fruition and is shown to be of practical benefit to humans, *then* a surgeon's interest is piqued. By that time, of course, the same important research conclusions can also be found in *Time* or *Newsweek* and will make perfect sense to both the average neurosurgeon and his next-door neighbor.

I wonder, then, if my gerbil work had made it into one of the top journals, how many people would have actually read it? Consider that there are about four thousand neurosurgeons in the United States. Most probably subscribe to the major journals but not all will actually crack open every journal. Of those who do, a large portion would gloss over the laboratory section, probably skipping my article altogether, especially as their eyes hit the words "gerbil" and "amiloride."

Then, of the subset of neurosurgeon-readers who would actually focus on the laboratory articles, only the ones interested in stroke would pay attention to mine. Of those specifically interested in laboratory stroke work, only some would actually read the whole article. Most would probably just skim the abstract. How many human beings, then, would have actually read my paper? I'd rather not know the answer, but at least I had the title of my project on my CV in time for the interviews.

FIVE

Culture

Before I became a neurosurgeon, I thought I knew what a neurosurgeon was supposed to be like. The ideal neurosurgeon was a James Bond type of character: calm, cool under pressure, precise in his actions, and culturally refined. I'm not sure where "culturally refined" came from, but as soon as I became a neurosurgery resident myself, I dropped it from my list of expectations. The social isolation can put a strain on refinement.

By chance, though, I have come across a few surgeons who reflected my preconceived notions quite nicely. The one who fit the bill most accurately was a general surgeon I worked with once, early on in my general surgery internship. A patient arrived in the ER late one evening with serious bleeding into the abdomen. The chief resident and I called the attending surgeon in for emergency surgery.

The man who arrived just happened to be British, with a true James Bond accent and the accompanying air of refinement.

He arrived quickly—but calmly—from a formal party wearing a well-tailored suit, and wasted no time in changing into his scrubs. In surgery, he was slick, efficient, and no-nonsense. He was in the guy's belly in no time. In rapid sequence, he isolated the bleeding vessel, repaired the injury, and left me and the chief resident to close the case. (I imagined a quick getaway, back to the party, in his Bond car.) Everything was effortless, including the politeness he maintained throughout, even toward the scrub nurse who had struggled to keep up the pace.

Upon leaving the OR table, he looked down at his shoes. They were speckled with blood. "Damn. I just brought these back from Italy." He walked out of the room, shaking his head. Saving the guy's life was no big deal for him: all in a day's work. The blood-speckled Italian leather was relatively more distressing. As a young intern—my first month on the job—I found that intriguing.

I should say a word here to clarify the stages of the training process. Residency is the period of time when a newly minted M.D. is in training to become a certain type of physician, such as a pediatrician, surgeon, or radiologist. Depending on the specialty, this can last from three to seven years, immediately following the four years of medical school. The very first year of residency is referred to as the internship. In surgical specialties, this intern year involves rotating through a variety of different surgical specialties.

The terms "junior" and "senior" resident are somewhat variable. In my residency program, a resident was a "junior" in years two and three, a "senior" in years four, five and six, and a "chief" in the seventh and final year. In some fields, in which further subspecialization is desired—pediatric neurosurgery, neuroradiology, transplant surgery, for example—additional training is required *after* residency in the form of a fellowship. A physician is then finally "in practice" after the completion of all this training.

As a junior resident, it became obvious to me that different neurosurgeons respond to stress in different ways. Many do remain quite calm, but not all are James Bond. I have witnessed temper tantrums, high-decibel yelling, and even the type of foot stomping you might expect from unruly children. I have observed instruments being flung to the floor. I have watched nurses flee from the room, scared to return. These behaviors certainly aren't refined or worthy of imitation, but they can at least offer amusement, as long as you remain a detached observer.

Unfortunately, a hot temper does tend to get the OR staff to spring into action, albeit begrudgingly and bitterly, and the fact that it works can further reinforce the bad behavior. Problems can arise, though, when a hot-tempered surgeon exhibits such behavior outside of the OR, like in a bank or a supermarket. He may be regarded as unstable. Although he may feel like a king in the OR and the hospital hallways, nobody outside of this isolated world recognizes or acknowledges his royalty. (His office staff may treat him royally, too, but this is an even smaller world.)

These days, in this era of political correctness, the worst offenders can actually be threatened with forced time off and sensitivity training. This fate befell one foreign-born surgeon who yelled at a nurse, in awkward English: "I cut your face!"

Neurosurgeons tend to be competitive in nature. This is a double-edged sword. As I mentioned before, neurosurgeons were the kids who always got a chair in musical chairs. I remember attending a residents' luncheon sponsored by the company that supplied us with our surgical drills. In order to lure us into the conference room so they could show off their latest drill bits, the company representatives (the "reps") had a nice buffet set up on one side of the room and a series of sheep heads lined up on the other side. We did not consider this strange. Food was necessary to get our attention, and the sheep heads were necessary to allow us to try out the drill bits.

To further pique our interest, they organized a clever competition for us. The resident who could create a standardized bone flap the fastest using their new drill bit would win a PDA device, or personal digital assistant. These devices were still new at the time, not yet mainstream, and were expensive, especially for a resident. (We all knew that the young well-groomed reps serving us lunch earned a lot more than we did, but we preferred not to dwell on such inequities.) Everyone lined up after finishing lunch and the reps presented the heads, one by one, and stood by with a timer. I won. My win did not sit well with one of my colleagues, though, who demanded an extra head. He drilled two additional bone flaps and was able to beat my time. Onlookers cited the well-known "practice effect" and discounted his time. He sulked. I took my PDA device home and used it for a few days before storing it in my closet. My white coat was already weighted down enough. Plus, my index cards worked just fine.

Although the culture of neurosurgery breeds certain collective traits, like confidence and a competitive nature, other traits are more variable. Take, for example, neatness. Here is a question to ponder: If you're trying to decide between two neurosurgeons with equally good reputations, do you go with the guy with a neat and organized office or the one with papers and charts strewn everywhere? Maybe the neat one has too much time on his hands and the messy one is in greater demand—no time for neatness. Or, maybe the neat guy appreciates form and the messy guy cares only for function. Although a sloppy desk (or manner of dress, for that matter) does not necessarily translate into messy surgery, the question would certainly enter my mind, if I were the patient.

My mother is quick to report her dismay at the clothing of various physicians she has seen. She was startled to notice dirty sneakers on one, and a simple polo shirt ("No white coat! No jacket!") on another. I was a bit embarrassed to learn that she did not reserve her comments for me, but was up front with her doctors as well. Although professional attire at work is still the norm among physicians, the trend toward more casual wear seems to have come about with the wide acceptance of the term "service provider" in medicine, lumping us together with the providers of all other sorts of services out there. (Fries with your craniotomy?)

There is no correct answer as to whether or not neatness counts for much, but, all other things being equal, I would prefer the neat surgeon if I were the patient. Neurosurgery is not plastic surgery, but I would prefer a surgeon with a concern for aesthetics. Take, for example, the issue of the hair shave. From the surgeon's standpoint, shaving a patient's hair prior to brain surgery is the most minor of concerns, understandably. The hair is just in the way of the scalp, which is in the way of the skull, which is in the way of the brain. Years ago, neurosurgeons would routinely shave the entire head, thinking that this was necessary to control infection. Over time, the extent of the shave lessened to include only a large patch around the incision, which is more efficient. Many patients, though, have told me that as long as a large patch is coming off, their whole head might as well be shaved. They have a point here.

More recently, restricting the shave to only a minimal strip of hair along the incision line has become popular among neurosurgeons who care about the hair issue (more commonly, I have found, the ones with hair themselves). This approach is especially gratifying for patients with longer hair, as the incision can be hidden entirely. These patients can go out in public without wearing a hat or broadcasting to the world that they just had brain surgery. I remember seeing a photo of Elizabeth Taylor in a magazine after she underwent surgery for a benign brain tumor. All I could think was: Why did they shave her entire head? That's so old school.

I met a girl who was a high school senior and who developed seizures due to a benign brain tumor. In order to control her seizures, the tumor needed to come out. I was to assist the senior surgeon on the case. The girl was looking forward to starting college and confided in me that she had just begun dating a guy "with the greatest muscles." She was equally concerned about the fate of her long hair as she was about the risks of surgery. The senior (bald) surgeon, however, was in the habit of shaving a large patch and his routine was set. Her hair was doomed and she knew it.

As the patient was being put under anesthesia, I pleaded with the senior surgeon. "Just let me do the hair shave. Don't worry about it. She won't get infected. You can blame me if she does." He didn't like the sound of it but he gave in to my desperate tone. He had to leave the room, though. He didn't want to watch. I shaved a thin strip of hair along the exact path of the proposed incision (a curved "reverse question mark" incision that we often use on the side of the head, just above the ear). I scrubbed her scalp and hair with an antibiotic solution and stapled the sterile drapes along the shaved edges to keep the hair out of the operative field.

The case went smoothly. In the ICU, after the patient was fully awake, she wanted the full report. She asked about her hair. I surprised her with the news and she was ecstatic. In fact, I have never seen a patient happier after waking up from brain surgery. She couldn't wait for her well-built boyfriend to see her. (Her delight made me wonder— but only for a moment, before I considered the ethical implications— if we should play a similar trick on other patients, setting up the expectation of a bald patch, but then surprising them with a full head of hair.) Not only that, but her seizures remained under control, too, which was something we all cared about, hair or no hair.

Neurosurgery requires a delicate balance between fearlessness and caution. As residents, we have to be willing to push ourselves to take the next step, even if our confidence level is not one hundred percent. Otherwise, we won't go very far. On the other hand, I fear the resident who forges ahead with brazen overconfidence. Some are tempted to do this in an effort to impress the "attending" (attending surgeon) when he walks into the OR, hoping to provoke a comment like "Wow. You're under the 'scope (microscope) already. I'd better get in there before you finish the case!"

This leads me to an awkward admission: in the training of a neurosurgeon, the level of supervision can be variable. Keeping a patient safe depends on the judgment of both the attending and the resident: the attending on knowing how much to trust the resident's skills and the resident on knowing the limits of those skills. Believe it or not, this very human system is much safer than it sounds. Neurosurgeons are generally intelligent individuals who exercise good judgment (in the OR, at least). As a result, the norm is good patient care and the turning out of a steady stream of well-trained neurosurgeons.

When I was a junior resident, after I had performed a history and physical on a patient scheduled to undergo surgery, the woman turned to me with the sweetest voice she could muster and said: "Oh, and by the way, dear, I don't want any residents involved in my operation." Although this may seem like a simple request, akin to asking for a private room, such requests were very rarely heeded at our institution. Our chairman, who was to perform the operation, explained to her that the numbers we quote regarding risk and outcomes are based on our tried-and-true routine perfected over the years at our teaching institution. This routine involved a team, and the team included residents. He did not like to deviate from routine. In general, it is wise to avoid deviating from the routine in surgery. Every surgeon can tell you about a mishap that occurred when a VIP was treated differently from everyone else.

If avoiding residents is critical to you, a private, nonteaching hospital is always an option. At these hospitals, surgeons usually operate with a variety of non-M.D. surgical assistants rather than residents. In addition, the surgeons in those hospitals are more likely to have surgery as their sole focus. Many academic neurosurgeons have multiple roles: surgeon, teacher, researcher, committee member. They often don't have time to spend the entirety of each case in the

OR, from start to finish. The residents will do a portion of the work while they parallel-process on other tasks. Some neurosurgeons are very hands-on and will stay in the room for nearly the entire case. Others will flit from room to room or between room and office, attending to the case when needed.

At the extreme of laissez-faire supervision is the government-run VA system (the Veterans Administration hospitals). Although supervision is always around the corner, and decent quality health care is the rule, the residents tend to run the show. At my training program, the affectionate term for our VA hospital was the "Va Spa," a sarcastic reference to the humble and somewhat depressing standard-issue municipal edifice. The expansive, oversized room that housed dozens of patients in long rows was nicknamed "the ballroom." On the door of our small single-room neurosurgery office at the VA, one of the residents scrawled, "Resistance is Futile." Common knowledge dictated that it was no use trying to change the system at the VA; better just to live with the quirks and inefficiencies. Physicians with Republican leanings are quick to point out that this is what a broader government-run health care system would be like. We'd get the care we needed, but we'd all have to lower our expectations a bit, and take a number.

Even though neurosurgery is a small specialty, you can't lump all neurosurgeons together. Some neurosurgeons tend toward the more brainy, nerdy end of the spectrum, whereas others are more of the jock or frat boy variety. Some love to spend all day in the OR whereas others prefer to spend extra hours in the lab. A few superstars do it all, via liberal delegation, collaboration, and little sleep.

Such distinctions can introduce a funny quirk into the academic system: a lay person might assume that a surgeon with his name on the greatest number of papers, or the one with his name in the *New England Journal of Medicine*, is *the* guy to go to for a certain type of surgery. While this certainly may be the case, and I don't discourage this type of thinking outright, the reality could also be that this is the guy who spends far more time in the lab than in the OR. A great mind for science and great hands do not necessarily go together.

Some renegade neurosurgeons have taken on research projects that have gone down in history—as much for inciting ethical debate as for pushing the envelope. In addition to being inspired by Harvey Cushing, my husband and I were also influenced by another neurosurgeon from Cleveland, but a living one we actually got to meet: Robert White. White became famous for many things, but perhaps most of all for his head transplant work, which he preferred to call a "body transplant." He actually performed a head transplant on a primate in 1970, connecting all the blood vessels required to perfuse the new head. The transplanted head was fully functional, but, with no way to connect the two ends of the spinal cord, the new being was quadriplegic. However, White envisioned that this experiment, however macabre, might someday pave the way for extending the lives of people with healthy brains but deteriorating bodies, such as those who are already quadriplegic or nearly so.

White kept popping into our lives at various times, sometimes unbeknownst to him. My father knew him from years earlier and set up a meeting—all four of us—when Andrew and I were still college students. We checked out the historic transplant lab. A year or so later, Andrew and I spent a summer month in Spain together. While staying in a rundown apartment in Valencia, we had a knock on the door one day. An animal rights activist was handing out flyers. It featured animal abuse "horror stories," including a description of White's work and the lab we had seen. We laughed, not at the purported abuse, but at the fact that our Cleveland buddy had made it to Valencia, to our little apartment.

Years later, White was nice enough to attend our wedding. And, upon buying our first car together and driving it off the lot, we just happened to turn the radio on to the sound of White's voice on a talk show. White has become legendary, especially in Cleveland, where he is perhaps best known not for his research, but for his connection to the late Pope John Paul II, for whom he served as a medical adviser. My father—a general surgeon who told me long ago that neurosurgeons are known for being megalomaniacs—likes to tell the following joke that circulated around Cleveland at one point. The joke goes: A crowd is gathered at the Vatican, with everyone looking up at the balcony, where two men can be seen. One random onlooker says to another: "Who is that guy standing next to Dr. White?"

You might think that neurosurgery as a profession is specialized enough, but we've managed to break it down into even smaller bits. And, as with any culture, stereotypes abound, some fair and some unfair. Vascular neurosurgeons, for example, are the "cowboys" whose lives revolve around aneurysms and other blood vessel–related brain abnormalities. Aneurysms are weakened bubbles on blood vessels that can burst, sometimes leaving neurological devastation in their wake. Emergencies are the norm. The vascular guys are known for having the worst lifestyle. Aneurysms seem to present themselves at the most inopportune times, like Friday evening or Thanksgiving afternoon. They often require a trip to the OR to prevent a rebleed. The spouses and children of these cowboys may file for neglect. On the bright side, vascular neurosurgeons do some of the greatest cases—intricate and technically demanding—the kinds of cases that attract young medical students to the field in the first place.

Spine specialists are a different breed. They work with screws, rods, and bones. They may be mistaken for orthopedic surgeons (some of whom do spine surgery also, but only if they've undergone special fellowship training). They are courted by instrumentation companies that hope to entice them with new products, like modified screws or updated screwdrivers. Spine surgeons are the ones who usually bring in the most cash, to put it bluntly. There are a few simple reasons for this: (1) the population is aging, (2) aging spines can be painful, (3) there's a growing trend toward fusing painful, aging spines, and (4) fusions reimburse well (better than most brain operations).

Picture this. A haggard vascular neurosurgeon comes in at midnight, Friday night, to do an emergency aneurysm case on a deathly ill patient. He monitors the patient for a week or two in the ICU and another week or two on the floor. He fears the potentially devastating complications that can occur even several days after a flawless operation. He answers frantic phone calls in the middle of the night. He holds family conferences. He checks scan after scan. He will be compensated reasonably well for the operation itself, depending on the patient's insurance plan, assuming that the patient actually has insurance. (Otherwise, he works for free.) However, he receives nothing additional for all the work required after surgery.

The spine surgeon, on the other hand, performs an elective spine fusion on a healthy patient on a Monday morning (after his office staff has confirmed insurance coverage). He checks on the patient once a day for a few days in the hospital, sends the patient home, and receives a multiple of what his vascular colleague received. This disparity, although seemingly unfair, has become an entrenched part of our culture. I introduce this disparity not as a crude exposé of our finances, but because these issues are on our minds all too often. Ask almost any neurosurgeon.

Neurosurgeons come in many other varieties: pediatric, functional (for movement disorders such as Parkinson's disease), tumors, trauma, epilepsy, and peripheral nerves. A small subset of neurosurgeons focus on anything that requires surgery at the base of the skull. These "skull base" neurosurgeons are famous for their maximally invasive approaches. Their cases may take all day, sometimes extending into the next day. For the longest operations like large, complex tumors at the base of the brain, neurosurgeons sometimes work in shifts, so that one surgeon can leave to use the bathroom, eat, and explain to their spouse why they won't be home, while the other one works. To many young residents, this complex and demanding field can be quite attractive . . . at first.

Radiosurgeons are in a rarefied class of their own. They perform stereotactic radiosurgery, a slick noninvasive technique that involves focused radiation to treat brain lesions. Their patients are usually quite pleased because they avoided having their heads opened—the more traditional alternative. Their procedures are not performed in the OR, so they have certain liberties that other neurosurgeons cannot share, like drinking specialty coffees during their cases. They're often the smartest ones around, partly because they have time to read.

The wide spectrum of technical approaches in our field, from minimally invasive to maximally invasive, and even noninvasive, inspired one graduating resident to offer this tribute at his graduation speech: "I thank Dr. X for teaching me to operate through a keyhole, Dr. Y for teaching me to operate through a manhole, and Dr. Z for teaching me to operate through no hole." This quote has been passed down, year after year, at my training program and I can't even remember now who should get the credit.

Neurosurgeons of all types, nationwide and even worldwide, convene at our annual neurosurgery conventions. These meetings are designed to strengthen our social bonds and keep us up-to-date on the latest scientific advances (in that order). Our two major organizations, the AANS (American Association of Neurological Surgeons) and the CNS (Congress of Neurological Surgeons), each hold a separate meeting. Most neurosurgeons belong to both.

Each organization has it own corresponding journal. The AANS publishes the *Journal of Neurosurgery* and the CNS publishes *Neurosurgery*. Although both are equally well regarded, the younger *Neurosurgery* is more colorful and has more pictures, provoking the nickname "the cartoons" by its rival. Among the more senior neurosurgeons, the older journal is known as the "white journal," based on the cover, and the newer one is the "red journal," further reflecting the divergent emphasis on color.

My first annual meeting was in Chicago, when I was a junior resident. Finally, after hearing about all the big names in neurosurgery around the country, I was able to match faces with names. (Their faces are not exactly featured in *People* magazine; neurosurgery is a very small sea. Once the big names set foot outside of the convention center, they become relatively anonymous again, unless they forget to remove their convention badges.) As I went up escalators and walked down hallways, the senior residents would lean over and whisper to me: *The guy with the mustache is Spetzler . . . that older guy is Yasargil . . . the big guy over there is Rhoton.* These were the neurosurgeons typically asked to give the "How I Do It" talks, like "How I Do It: Giant Aneurysms," or that type of thing. The same talks tend to be featured year after year, which didn't dawn on me until my second or third meeting.

One of the most unusual aspects of these gatherings is the massive, open, sterile convention room filled with specialty vendors. I'm not talking about food vendors. I'm referring to companies that sell things like surgical instruments, medications, textbooks, and multimillion-dollar pieces of capital equipment. They vie for attention with colorful, educational displays and freebies such as pens, candy, customized Post-it notes, and squeeze toys in the shape of a brain. At some of the booths, you can try out the equipment. If you're so inclined, you can, for example, test the performance of a cautery device on a piece of raw steak. You can look at fine newsprint through the lens of the latest surgical microscope. You can peruse the gamut of OR tables that fold over, orgami-like, in various ways. These tables often feature live models in black leotards lying motionless, simulating a patient about to undergo spine surgery. The models are always women.

Residents get to eat well at these meetings, which is much appreciated in the context of a chronic hospital cafeteria diet. Dinners at the best restaurants are sponsored every night, either by our academic department or by a company representative. Expensive steakhouses are favorites. As a petite woman, I usually can't finish the entire piece of steak, but it never goes to waste. There is always at least one guy in the pack who is willing to wolf down the rest. Someone asked me once whether there were any particular advantages to being a woman in the male-dominated field of neurosurgery. I mentioned this one.

The scientific information at an annual meeting is presented in various formats. The studies deemed most important, like the results of large, multicenter trials on aneurysms, brain tumors, or spine surgery, are presented in the biggest rooms during exclusive times when nothing else is going on, so the maximum number of people can attend. Talks that are deemed important, but a little less important, are also granted time slots, but the talks are shorter and have to compete for attention among the other sessions. Everything else is presented in poster format. Apparently, most posters that are submitted are accepted; relatively few are rejected. This ensures maximum attendance, as a potential presenter might skip the meeting if his poster is rejected, leaving the organization with one less registration fee. I didn't know this when my first poster was accepted. I was thrilled that it had surfaced to the top of what I had envisioned was a highly competitive weeding-out process.

A poster always presents a dilemma: What do you do with it when the meeting is over? Some departments are willing to devote some hallway space to it, but if not, it's a toss-up whether or not to haul the unwieldy tube back home on the plane. After the devotion of so many hours, it seems a waste to throw it out. That's probably the best option, though, because few spouses would encourage its display in the living room or bedroom.

The slogan at my first annual meeting was "Winds of Change." It was displayed in large print on banners and on the cover of our meeting programs. The slogan was fitting, given the constant evolution of our profession and the host city of Chicago. The society president gave an uplifting talk at the opening of the meeting, reflecting on the recent unique "winds of change" in our profession. I was proud to be a new member of the society, sitting among thousands of other residents and neurosurgeons from around the world.

On the last day of the meeting, I walked through the length of the convention center on my way back to the hotel. I could see that a national hardware chain was preparing for their own annual meeting, scheduled to follow ours. As I walked, I noticed more and more name tags bearing the logo of the hardware chain, while the neurosurgery badges thinned out. They were taking over the space I had come to consider ours. Upon leaving the building, I saw their banner hanging proudly at the main entrance. I paused when I saw their slogan: "Winds of Change." I imagined their opening talk to be equally uplifting.

SIX

Routine

"Don't go into neurosurgery unless there's *absolutely nothing else* you could ever see yourself doing." In other words, unless you're fanatical about it, it's not worth the sacrifice. I received this advice from elders in the field, as did other medical students hoping to enter the specialty. For some, the advice triggers introspection, which is the whole point. For others, the gravity of the message and the solemnity of the delivery only enhance the aura of exclusivity surrounding the profession. For those starry-eyed medical students, the thought of joining a tribe of devoted and single-minded practitioners—a tribe that others are not passionate or qualified enough to join—makes the decision even easier. The warning is pure enticement.

I was in the more introspective camp. Unfortunately, though, the warning continued to haunt me even after I had made up my mind, extending my introspection through seven years of residency (and beyond). I am convinced that most neurosurgery residents question their decision at least once or twice during the prolonged training period. Regarding those who didn't—the ones who claimed an unwavering confidence in their career choice—my feelings alternated between suspicion and jealousy. Some seemed to have been born into the role. I, on the other hand, smiled and got through the training just as deftly, not so much because of a natural fit, but, at least partly, because of my innate tolerance (and, usually, fondness) for hard work. I took it in stride but I didn't always like it.

The decision to become a neurosurgeon places you on a track that runs, unabated, through a seven-year tunnel. This begins only after completion of the prerequisite four years of college and four years of medical school. This means that the average neurosurgeon is in his or her early thirties by the time the "real job" begins as a fully fledged surgeon with a decent salary and independent decision-making. If you tack on a one- or two-year fellowship (or worse—a Ph.D.), then you're talking mid- or even late-thirties. (By that time, a good deal of interest has piled up on student loans.) At the end of the tunnel, the formerly undifferentiated M.D. emerges as an exquisitely superspecialized neurosurgeon, squinting at the rest of the world—a rare animal dominating a small niche within the ecosystem of medicine. At that point, you feel unqualified to do anything else, even if you had any lingering thoughts about a career change. It pays to listen to the elders before entering the tunnel in the first place.

Neurosurgery is marked by labor-intensive routine sprinkled with brief highs. The highs keep us going, so we hope they aren't too brief or spaced too far apart. Popular portrayal of surgery on television usually focuses on these highs: saving a life, removing an ugly tumor from an otherwise young healthy brain, separating conjoined twins fused at the head. The laborious routine is far more representative, but not as enticing. (Only a very small handful of neurosurgeons, by the way, have ever separated conjoined twins, including one who wrote a book entitled *Gifted Hands*, about himself.)

In the traditional culture of neurosurgery, the puritanical work ethic reigns supreme. Our routines require long hours, especially during residency. Because we're stuck with these long hours, we turn the long hours themselves into a source of collective pride. We tend to make fun of the "softer" specialists who never get called to the ER and who make it home in time to help their spouses chop vegetables for dinner.

One morning during residency, at our usual post-rounds seven a.m. team breakfast (after arriving for duty a couple hours earlier), my colleagues and I were joking around—and fantasizing—about the life of dermatology residents. One of the more contentious members of our group decided to put them to the test. He called the hospital operator and requested a stat overhead page to the dermatology resident on call.

Five seconds later, to the amusement of the entire hospital, the unprecedented page came through: "Stat page, dermatology resident on call, 4072. Stat page, dermatology resident on call, 4072." We erupted into uncontrolled laughter. Assuming that the dermatology resident was still at home (making pancakes?), we laughed even harder when he actually called back, promptly. According to my colleague, the guy on the other end of the line displayed a mixture of confusion and excitement in his voice, and was genuinely disappointed to learn that there was no dermatologic emergency.

Although team camaraderie, in addition to the brief and scattered highs, helps get us through the routine, the routine is still often a chore. Consider what a typical day is like in the life of a junior neurosurgery resident, based on my experience from the not-so-distant past.

I wake up at 4:30 a.m. in order to make it to the hospital by about 5:15. After a very brief shower, I get dressed in a fresh pair of scrubs and apply undereye concealer to try to mask the dark circles. I don't waste time with any other makeup. No blow-drying or other styling, of course. I'm satisfied with the plainest of looks in the interest of efficiency.

The one nice thing about getting to the hospital so early is that there are plenty of spots in the parking garage. I jog the short distance across the street between the garage and the hospital. I'm always rushing. I've never liked the feeling of having to rush, but there's no good alternative. We have to be on time. Before joining the rest of the residents at six, I need to get the overnight update on my patients. I need to check their vitals and their labs, do a quick neurological exam, leave notes and orders in the chart, and speak to the nurses. As a junior resident, I am in charge of the head and spine injury patients in the trauma ICU. Sometimes I come in ten minutes later or earlier, depending on how many patients were in the unit the night before. Coming in a little later, though, can be risky if a new "stealth admit" came in overnight. Then I really have to rush, as being late for rounds is strongly discouraged. It can screw up the entire team.

I have five patients to see on this particular morning, which may not sound like a lot, but it is when you consider the complexity of their care and the fact that I have less than ten minutes to spend on each patient. (On principle, I'm not willing to get up much before 4:30 a.m.—although a few residents do—in case you're wondering why I don't just come in earlier.) I see that there is one new guy from last night: a typical motorcycle crash, no helmet. I'll need to spend a couple extra minutes on him and skimp a little on another patient who's more of a "chronic player" (a patient who's been in the hospital a long time).

As soon as I enter the unit, two nurses start firing off questions about their particular charges, even though they know that I'll be going through each patient, systematically, one at a time. Yes, go ahead and repeat the potassium on that guy and, fine, I'll unclog the drain on that guy when I get to him. Some residents are cordial in answering urgent requests on nonurgent matters and others bark at the nurses for poor prioritizing. It can depend on how much sleep we got the night before.

I walk to the nurses' station to gather the charts. I notice greasy pizza boxes containing half-eaten crusts from the night before and a large open bag of Doritos that has probably seen more recent activity. (When you round in the ICU the morning after a holiday, a cold leftover potluck of casseroles, bratwurst, and fried chicken usually awaits.) Mildly nauseated, I walk to the end of the unit and start with the first patient.

I get bogged down on this guy because his exam is not quite as good as it was the night before, when I last examined him on evening rounds. The change could easily be due to the fever he developed this morning, but I can't make any assumptions. I look at his nurse and she fears what's coming: he needs to go down for a scan of his head. I call radiology and she starts to "pack him up" for the ride down, which requires disconnecting all of his tubes and monitors and getting a respiratory therapist to "bag" him (pump in artificial breaths by squeezing a large plastic grenade-shaped device connected to the tube in his trachea) on the way to the scanner while he's off the ventilator. It's a real pain, and I get all the usual grumbling. Although I can't say that the scan truly needs to be done stat, I know that if it's not done now, then no one will look at it until we're out of the OR much later in the day.

Of equal importance in my mind in placing the order "stat" is the fact that aggressive action is a highly valued trait in a neurosurgical resident. My early morning order allows me to announce on team rounds: "His exam was a little worse, so I sent him down for a stat scan. I've already checked it out and it's fine." Otherwise, if the problem isn't resolved, the chief is likely to say, "So what did you do about it?" and consider you weak. A lesson learned early on is that a sin of commission is better than a sin of omission. Better to do too much than too little. If you appear weak or indecisive, people will walk all over you.

I check my watch and rush through the next four patients before running down to meet the team. I don't have time to reflect on the fact that none of my patients are conscious enough to speak to me. I don't know them as people, but rather as complex data sets and problem lists. As they start to awaken, they slowly emerge as individuals. At that point, though, they're transferred out of the ICU and replaced by new data sets and problem lists.

The team gathers around the glow of a large computer monitor in the dark, windowless radiology department to review all the scans performed overnight. Our group consists of an intern, two junior residents, two senior residents, and two chief residents. One of the residents is still wearing his red outdoor fleece, which means he was running late and didn't have time to get to the office to change into his white coat. (We heckle him about another "fleece day.") Everyone else is wearing scrubs and white coats. According to official hospital rules, we're supposed to wear our "street clothes" to the hospital and change into scrubs in the locker room. That's a precious waste of time, though, at five a.m. Everyone just takes their scrubs home with them, in defiance of a rule made by someone with more time on his hands.

We have about ten minutes to examine the new scans, and the senior resident who was on call overnight "drives," selecting the appropriate images on the appropriate patients from the computer. (In the modern era, films are no longer printed out and read on a light board.) The review is perfunctory and solemn until we get to the scan of a patient who required placement of an emergency ventriculostomy, a tube placed in the brain to relieve pressure.

One of the chiefs pipes up: "Wow! Who put that one in?" Everyone laughs heartily as the senior resident driving the computer lays claim to the job: "It's working just fine, so I don't even want to hear it from you guys." These drains are placed freehand, often under duress, through a small hole in the skull, aiming for a certain spot deep within the fluid spaces of the brain. There are numerous ways in which the tip of the catheter can end up a little off from the desired target. This is typically of no consequence to the patient, so the images become a fertile but largely harmless source of ribbing among the residents. The ridicule is heightened, though, if the case does become complicated and the scan is reviewed on the big screen in our conference room, in front of the entire department. (These drains can be not only a source of social embarrassment, but also the cause of interrupted sleep when they become clogged in the middle of the night and you are called to the ICU to flush out whatever is clogging it, like small bits of brain—affectionately known as "brain guppies" in our program—or blood.)

We rush en masse from radiology up to the first ICU that we need to cover, and the resident responsible for this group of patients presents them in the ritualized, efficient manner that everyone is accustomed to. If the resident deviates too much from this form, the chief is likely to demand a course correction. Sloppiness and inefficiency are not tolerated. Everyone else on the team needs to take notes for themselves, and a disorganized presentation screws everything up on our sheets.

Here's an example of a typical presentation on rounds, in written form as would be found in the hospital chart: "Joe Blow. PBD 8, POD 7, SAH, L. PComm. Levaquin #4/7 (lung?). AVSS, Tmax 38.2, SBP 120s–140s. I/O 4.2/3.1. D5 1/2 NS @ 100. Na 140, Hct 35. AA Ox2 ("1998"). PERRL. Speech normal. No drift. Strength 5/5. A/P: Neuro stable; Tx 5G, d/c Foley, cont. nimodipine." The presentation takes less than one minute and the chief is then free to add to or subtract from the plan, to praise or to criticize. The chief's other task is to keep the team moving. There are too many other patients to see, and we can't get bogged down on any particular one.

This brief presentation supplies the entire team with all the pertinent details required to cover the patient. Equally important, it allows any resident to speak intelligently to any attending about any patient when accosted in the hallway. (The sight of an attending usually prompts a reach into the white coat pocket for our notes from the morning.) In the case of Joe Blow, any resident can be confident in the knowledge that the patient suffered a bleed from an aneurysm located on the posterior communicating artery in the brain. The bleed occurred eight days ago, the surgery was seven days ago, and he's been on antibiotics for four days (with plans for seven days), for presumed pneumonia. He had no fevers over the past twenty-four hours, his blood pressure is fine, and his fluid balance is fine. He's still on intravenous fluids. His pertinent lab results—sodium and blood count—are normal. He is alert and oriented to his own name and place, but not to the year (he thinks it's 1998). His pupils respond to light, equally. His strength is normal. The overall assessment is that he is neurologically stable. The plan is to transfer him from the ICU to the step-down unit, to remove the catheter from his bladder, and to keep him on a medication to help prevent spasm of the vessels in his brain.

Despite our best efforts, there are numerous threats, lurking around every corner, that can potentially derail our efficient morning routine. Sometimes the threats are minor, like a chart falling apart, scattering dozens of pages across the floor. (The intern or the nurse will scramble to put it back together, while the patient is temporarily passed over. Given our penchant for superstition, someone might mention that the chart explosion foreshadows a poor prognosis for that patient.) More serious, a head injury could arrive in the ER during rounds, which means that someone has to "fall on the sword," peel away from formation, and run down to check it out (and maybe have to skip breakfast, leaving him hungry all day in the OR). A nurse from the preoperative holding area may call about a missing surgical consent form. This one is particularly annoying. It's purely a paperwork snafu, but it demands our attention because it threatens to delay an operation. The guy lowest on the totem pole is usually sent down to sort it out.

If all goes well, the entire team makes it up to the cafeteria for a working breakfast. That's when we receive the update from the intern regarding the floor patients. Patients who have made it to the floor are usually pretty stable and so we trust the intern to look after them. Vexing social issues tend to be more common than medical issues: Mrs. Y's family doesn't want to take her home and they don't like any of the nursing homes they've seen; Mr. X's wife is upset that Dr. Z hasn't been by to speak to her in two days; Mrs. Q is still complaining of pain but her insurance won't cover another day here. In general, we can't focus much on these things as residents. They are too remote from the life-and-death end of the spectrum and we don't have much time for them. They wear on us, though, like water on a stone.

At the end of breakfast, the critical discussion of the day takes place: Which residents will be assigned to which operations? This largely dictates how the rest of our day will go. Seniority rules, with chiefs granted their first choice. Many factors play into the decision: which attending is involved, how technically challenging or "fun" the operation will be, whether it's a "fresh" case or a "redo" case, and whether the operation is routine or uncommon. One of my chiefs would display his seniority as a royal might display royalty. He would look over the schedule and announce to the rest of us, aristocratically: "Today I will be in room seven with a fresh microvascular decompression, followed by a light lunch, and I'll finish off my day with a nice microdiscectomy." The same chief was famous for telling us at breakfast: "If you need me, I'll be in the pons" (a part of the brain stem).

The more junior residents came up with clever ways of enforcing fairness in their more meager choices, the leftovers. A common method was to use a special trump card that was passed between residents of the same rank, usually kept in the breast pocket of the white coat. During one year, a photocopy of our chairman's face was taped to the card to give it an air of official sanction. The card was more often used to avoid a case that nobody wanted to do rather than to select a choice case. It was handed between residents, always with a touch of drama and ceremony.

By the time breakfast ends at seven-thirty (just in time for the OR), we have discussed every last patient. Given the natural ebb and flow of disease, trauma, and our attendings' national meeting schedules, our neurosurgery service can consist of as few as twenty-five patients and as many as seventy. The census typically hovers around the fifties or so. Central to keeping everything organized is "the list": a continually updated computerized roster of all the neurosurgery patients in the hospital. The list is so critical to keeping everything straight that any perturbations can cause widespread grief, and even panic, across the entire team. When the computer system is down altogether, and lists cannot be printed for morning rounds, the f-word is bounced around liberally among all the residents. Everyone is then forced to write all the names by hand, and the mood quickly sours.

One of the worst fates that can befall an individual resident is to lose the list while on call. This can ruin your night. It's not that the computerized list is irreplaceable—it can be reprinted—it's that the notes scribbled underneath each name serve as critical reminders of who the patients are and of the numerous tasks required of you while on call. If you lose the list, you have to try to recall who was having which scans, who needed certain labs checked, and who needed a spinal tap. Each resident knows a handful of the patients very well, but the rest are known only secondhand, from the brief presentations on rounds.

A resident can highlight the tasks on the list in many ways: putting little boxes in front of each task, lining them up in a column along the right-hand margin, using a different pen color, or circling them. It was the rare resident who had no consistent system at all, and I always worried about that resident. Critical hallmarks of a model resident are consistency and obsessive-compulsive attention to detail. A couple guys I knew, though, were a bit too obsessive, faithfully recording every last normal potassium level. I worried about them, too.

When you're left alone in the evening with your to-do list, triage is key. You want to complete the essential tasks first, in case you end up in the OR for a few hours and can't get to the rest of the list until early morning. Here's what might be on a typical to-do list for one evening as a resident on call: check four scans, check three lab results, reexamine a postoperative patient and call the attending neurosurgeon with the details, do a spinal tap on one of the trauma patients with a fever, talk to an irate family member on the floor, and switch a medication dosage. As these tasks are attended to, other things pop up at random: urgent pages need to be answered and emergencies in the ER need to be seen. In between all this, you try to grab dinner.

Sometimes you can get a little sleep while on call, and sometimes not. Because I was the first and only female resident in my residency program (until the second woman was accepted a couple years later), the call room had a decidedly masculine feel. I still remember the "Tall Cool Red One" beer poster that hung on the wall, featuring a thin attractive redheaded woman in a bikini. The room was small, but there was enough wall space for two other posters, also beer posters, ones with apparently less memorable slogans. The other notable feature in the room was the dent in the wall near the phone. The phone hung just above head level on the wall next to the bottom bunk bed. The dent's origin was easy to envision, as it appeared to be created by the slamming of the receiver end into the wall. I mentioned earlier that urgent questions about nonurgent matters can try a resident's patience. This is especially true at three a.m. when awakened from a stolen hour of sleep.

In the morning, you begin the same routine again. The post-call morning is different, though, because it's more difficult to stay awake and you haven't showered (a ponytail day for me; many of the guys had crew cuts). In addition to being tired and dirty, you're also open to critique on whatever actions you took (or didn't take) overnight, combining a sleep-deprived irritability with a vulnerability to attack.

Arriving back home late Sunday morning following a sleepless Saturday night on call is absolutely heavenly. I would be content to sleep the day away, but my husband, Andrew, saw that as a declaration of defeat, another admission that we led highly abnormal lives. One Sunday morning, after I had spent the night covering the pediatric service and he had spent the night covering the adult service, we returned to our sun-filled town house. Neither of us had slept. I ran to the bedroom, shut the blinds, and jumped in bed. He reopened the blinds and announced that we were going to go on a hike, to enjoy the outdoors: "Don't be lame. We have to have some kind of a life," he implored.

I gave in but secretly promised myself that I would go to bed really early, like immediately following dinner, if not during dinner. We drove an hour away—not too smart in our condition, I admit in retrospect—to one of our favorite hiking areas alongside a whitewater river. We pulled into the parking lot, where rugged outdoorsmen were lifting kayaks off the roofs of their cars and muscular hikers were loading up their backpacks.

Andrew parked the car and I begged: "Let me just lean my seat back for a minute. I'm not quite ready to get going. Let's relax for a second." I could barely keep my eyes open, and Andrew gave in, leaning his seat back, too. It was nice outside. He cracked the windows. With the sun and the breeze, we both fell asleep, instantly. Several hours later, as the temperature in the car dipped with the setting sun, we woke up, put our seats back up, and headed home. We hadn't set foot outside the car, but at least we had put on our hiking shoes and made the effort to get out of the house.

When the hospital became my home during the worst stretches of my residency, even the smallest forays out into society got me all excited. I absolutely loved going to the grocery store. Here were all sorts of people around me, not immediately worried about their health, exercising their freedom to walk up and down the aisles with no care other than what type of ice cream they wanted to buy. That was beautiful.

Being outside of the hospital did have its downsides, though. The fact that my job tended so strongly toward the serious (and often depressing) made me a little intolerant of people who voiced excessively trivial concerns. To this day, I still get mildly annoyed by people who are frazzled and tortured by the least important decisions: Pulp or no pulp? Skim or 2 percent? I feel like butting in: "Look. You're not deciding whether or not to pull the plug. Lighten up."

Going to a movie was a treat, but I often had trouble staying awake. When Andrew and I go to rent a movie now, he often has to remind me: "We saw that one, remember? Oh, that's one that you slept through." Perhaps my greatest weakness as a neurosurgery resident was that I love getting at least seven hours of sleep, and eight is perfect. I wish I could say that I was just fine with four or five, but that would be lying.

Second thoughts about the career choice tend to crop up several months into the junior year, when you realize that the routine is unremitting, you're chronically sleep-deprived, and the highs seem a bit further apart than expected. But once you've made it that far, you figure you should probably just keep going, just like everyone else ahead of you is doing. When you're still near the start of the tunnel, it is always insightful to hear from those who have already made it out. I saved an e-mail message that got tacked to a bulletin board in the residents' office. It contained advice to the junior residents, from a recent neurosurgery resident graduate. The cynical tone is typical of the culture:

"Remember to use the Heisenberg Uncertainty Principle. Keep moving to avoid detection by the Boss. Create the Illusion of Industry. Show up at Key Events (Resident Dinners, M&M). Sit in the front row. Try to do well on the Boards because it gives you breathing room to goof off afterward. Don't volunteer any information. Fight the temptation to offer your opinion. You will regret saying anything that can and will be used against you. Keep your mouth shut. Do not become overly familiar with Dr. X... Remember, you probably won't be doing aneurysms in the community anyway. Too much work for not enough cash. You need the ancillary services available too, and most hospitals do not have interventionalists. Dr. *Y* and Dr. *Z* are the Most Powerful and should be smoothed. That's my lecture for the day."

Now that I have emerged from the tunnel myself, I look back on this lecture and wonder what I would add (or correct; the reference to the Heisenberg uncertainty principle may not be quite right). The advice is fairly comprehensive as it stands, but I can at least offer this additional humble, concrete pearl of wisdom: resist the temptation to eat from the hospital vending machines and the nurses' stations on a daily basis. The stress and the erratic schedule of residency threaten to derail even the most balanced physiology. You have to work to avoid entropy, or you end up in trouble. Residents who gain weight tend to fall behind the team in the stairwells and hallways, huffing and puffing, during the mad dash through morning rounds. They can't button up their white coats as comfortably. They break a sweat when everyone else remains cool.

Whereas some of the residents do remain principled in their eating and exercise habits (stealing away to what's known as the "ortho library," the weight room located in a forlorn corner of the hospital), others tend to pack on extra pounds in concert with the extra stress. Admittedly, I did slack a bit on the exercise, but I endeavored to maintain a reasonable trimness—perhaps, at least partly, inspired by the same "Tall Cool Red One" on the wall that probably should have offended me.

SEVEN

Evolution Through Blood

Two competing theories of evolution distinguish themselves by the gradations of change: gradual versus punctuated. According to the more traditional theory, evolution occurs slowly and steadily. The punctuated theory describes an uneven evolution. Based on fossil records, the notion here is that periods of relative stagnation are punctuated by more sudden and dramatic changes, propelling evolution forward at irregular intervals.

Through our training, neurosurgery residents evolve from lowly interns to fully fledged neurosurgeons. Based on my own experience, I believe this evolution is both gradual and punctuated. Although the learning curve continues unabated throughout training (and beyond), certain events push the process along at accelerated clips.

By the way, I have heard a few neurosurgeons refer to their own skill as a "gift," as if their ability to remove a tumor were granted—

fully formed—from above. This is more akin to the creationist theory and won't be covered here. I think such skills are learned, not granted by an abstract deity.

Bleeding is a simple but pervasive theme in neurosurgery, and one that sheds light on the punctuations in our evolution. There are at least two situations I can think of that constitute informal rites of passage in the training of a neurosurgeon: removing a life-threatening blood clot by yourself and controlling profuse, active bleeding by yourself. The *by yourself* part is key. The solitary nature of these acts accelerates the evolution of self-confidence above and beyond the comparatively straightforward acquisition of manual skills. It's one thing to act with confidence when someone is looking over your shoulder and can step in to assist (or bail you out); it's another to remain confident when no one else is scrubbed in with you. At this point, you may be thinking, "I don't want any trainees going through this rite of passage when *I'm* on the table!" but that's how surgeons come of age, and it's not as dangerous or as cavalier as it sounds. Our hands have been through all the motions before, and the necessary supervision is always around the corner.

Just to remind you, if you come to the ER of a teaching hospital in the middle of the night with a neurosurgical emergency, the resident on call will be the only one immediately available. Senior help can be called in from home, if needed. At nonteaching hospitals, there will be no resident. The staff neurosurgeon will have to be called in, which generally requires that he or she remain within about half an hour from the hospital when covering the ER (hopefully there's no traffic). So, if you are still squeamish about the concept of the surgeon-trainee, consider the fact that this extra half hour may be unkind to a brain already under duress. I would be grateful for the omnipresent and energetic trainee.

Returning to rites of passage, I am reminded of a historic coming-of-age ritual for boys entering adulthood in certain Native American tribes: for a prescribed period of time, they had to survive alone in the wilderness, gathering whatever they could find to eat, and relying on their own skills and instincts while awaiting a vision. (Com pare this to other coming-of-age rituals, like the bar mitzvah or first communion, especially when it comes to the options for filling the belly.) Although the personal evolution of a neurosurgery resident is not quite so ritualized, the punctuating events are equally compelling and memorable, and often call for a similar self-reliance.

Consider one particular night of punctuated evolution in my training. I was a junior resident, which means that it was my first fully dedicated year of neurosurgery. In our jargon, I could also say that I was a "PGY-2," or in my second postgraduate year. (The initial year after medical school is internship. Interns rotate through a number of surgical specialties—general, orthopedic, plastic, transplant, et cetera—and are not yet exclusively neuro-centric.) I was on call overnight and it was around midnight. At this hour I was, of course, the only neurosurgical representative in this large teaching hospital. My mentors were at home sleeping, hoping not to hear from me.

The ER paged me with a level one trauma alert (the most severe). They could tell, based on the paramedics' report called in from the ambulance, that a head injury was part of the picture. By the time I dropped whatever scut work I was doing and ran down there, the patient had arrived and the members of the trauma team were hovering like worker bees, performing their initial assessment and shouting out their findings to the nurse taking notes in the corner of the room. "Equal breath sounds!" Check. "Palpable radial pulse!" Check.

One of the first things you learn about managing trauma victims is to respect the "ABCs": airway, breathing, circulation. The neurosurgery resident shouldn't butt in to do a neurological assessment until the trauma team confirms breathing and a pulse. The brain is no good without blood flow or oxygen anyway, so we are willing to wait our turn at the edge of the trauma bay. Orthopedic surgeons need to wait in the hallway even longer. The brain takes precedence over broken bones.

The bummer here was that the patient had just received medication to paralyze him, moments before. He had been agitated and moving around too much, and could have been a danger to himself or others. And, it was difficult for the trauma team to assess the guy, flailing around like that. He had a breathing tube in place and the respiratory technician was at the head of the bed "bagging him." A paralytic medication paralyzes nearly all of the muscles in the body, including the ones involved in breathing.

In these situations, the only neurological function we can check is pupillary response (the tiny muscle that controls dilation and constriction of the iris is one that still works despite a paralytic). The other reflexes are pharmacologically suppressed, so there's really nothing else to test. I shined a penlight in his eyes and saw that both pupils constricted briskly and evenly. I added my two cents: "Pupils equal and reactive!" Good. At least he had that going for him.

We can also glean a few important facts, secondhand, by asking the paramedics what his neurological exam was like in the field. I walked out to the hallway and took one of the paramedics aside as he was packing up his equipment.

Apparently, the patient was an unrestrained driver in a motor vehicle accident and was ejected from the car. The paramedics found him on the side of the highway. Nice. If he had simply worn his seat belt, he would have been awake and telling his own story. (He wouldn't be telling me, though, because my services probably wouldn't have been necessary in the first place. I would have completed my menial tasks for the evening and headed off to bed, actually getting some sleep before going to the OR the following morning.) His wallet was recovered from the back pocket of his jeans. Based on his driver's license, we knew his name and that he was eighteen years old.

The paramedic told me more. Surprisingly, although our subject was unconscious in the sense that he never opened his eyes or verbalized anything beyond a moan, he was still "localizing" at the scene: in response to a strong pinch of the chest, he would reach up and try to swat the hand away. This demonstrated that he had a good deal of higher brain functioning. He was at least sophisticated enough to feel pain, localize it, and try to stop it. These types of movements are con sidered purposeful. The ability to follow commands ("show us two fingers," for example) is one step ahead of localizing. He wasn't able to achieve that level of sophistication.

We have a standardized grading scale for head injury patients, the Glasgow Coma Scale (or GCS). The exact details aren't necessary here, but suffice it to say that the scale has three components: eye opening, verbalization, and arm movement. We can make some rough predictions about long-term outcome based on a head injury victim's initial score at the scene or in the ER. Based on the para-medic's report, our patient scored a five out of six—not bad—in the last category. He was certainly not a lost cause.

Doctors often speak in a very matter-of-fact way during critical situations like this. They discuss the facts, the protocols, the proce-dures. An outsider might find this discourse cold and unfeeling, even bordering on cruel. After all, we're talking about an eighteen-year-old kid here. He's a senior in high school. He has parents, siblings, friends. He's part of a community. He has a future. Maybe he has a girlfriend. Was she in the car, too? The emotional impact is stagger-ing. He's not just a collection of clinical data.

Believe me, I have these thoughts, too. In fact, I can become plagued by these thoughts. During the first few moments in the ER, though, dwelling on them is distinctly unhelpful. In fact, anything that slows us down is not only unhelpful, but possibly even harmful. The ability to act quickly could mean the difference between this eighteen-year-old going to college or heading for a nursing home. We do have feelings, but our predominant thought in the heat of the moment is: we have the ability to help this guy; what do we need to do for him? The cool clinical demeanor reflects the practicality of getting a job done; it does not reflect a lack of humanity. Later, when his bodily organs are stable, we can reflect on him as a person, at least until the next ambulance arrives.

At this point, the trauma team needed to get him to the CT scanner, just a few steps down the hall, as soon as possible. (In busy trauma centers, it is nice to have a dedicated scanner right there in the ER.) The team was a machine. He was in the scanner within about ten minutes of hitting the door. After positioning him on the scanner table, everyone scrambled into the control room to watch the images as they appeared on the monitor, one by one.

A head CT starts at the base of the brain and works its way up to the top in serial slices. The first few images looked fine: no blood on the lowest cuts. The critical cuts through the temporal lobe, though, were a different story. Freshly clotted blood on a CT scan is bright white. There was plenty of bright white on these cuts. The patient had developed a large right-sided acute subdural hematoma, a blood clot that forms between the brain and inner surface of the skull, just underneath the dura, or outer lining of the brain. In this scenario, as ongoing bleeding expands the clot, the brain is pushed aside and can knuckle into the neighboring brain stem. Once the brain stem is threatened, life is threatened. A telltale sign of this knuckling, or herniation, is dilation of the pupil on the side of the clot. The nerve that controls pupillary response (the "third nerve" or "oculomotor nerve") runs just adjacent to the brain stem. His pupils had been fine just a few minutes prior.

I ran from the control room to the scanner. I opened his eyelids. His right pupil was now "blown"—our short way of saying fixed and dilated. It was huge and round and did not respond to light. This clot had expanded (and was expanding) before our eyes. My next thought was not really a thought but a knee jerk. "We're going to the OR." The "we," though, was really just "I." And, for the first time, I was rushing a guy to the OR by myself with absolutely no time to kill. If

I waited until senior supervision arrived, this eighteen-year-old was definitely not going to college.

I was ready for this. The neurosurgeon who taught me the ropes is a nationally renowned head injury guru and had taken me through all the steps before. He warned me that if someone is dying of a sub dural hematoma in the middle of the night, he doesn't want to hear my voice on the phone. He wants to hear the voice of the OR nurse relaying the message that I'm in the OR and the patient is on the table. So that's what he heard.

From the ER, we wheeled the patient into the elevator for a short ride up to the OR on the second floor. I killed the otherwise idle fifteen or twenty seconds in the elevator by shaving the right side of his head. We keep extra razors in the call room. One of my senior residents told me to keep one handy at all times in my white coat when I'm on call, for occasions just like this. Sometimes a razor will get passed, batonlike, in slow motion, from the resident finishing call to the resident starting call, as if in a relay.

In the OR, the anesthesia team kicked into high gear, receiving the handoff from the trauma team—another well-oiled machine. The OR nurses were flawless, too, unwrapping the instrument trays, wheeling the gas tank for the drill into place, setting out the sterile gowns and gloves. (They had done this a few times before.) I sprayed Betadine, a brown antiseptic solution, over the patient's freshly shaven scalp and got started. Few words were exchanged: ten-blade . . . retractor . . . drill . . . suction.

By the time my attending walked in, the clot was just on its way out. He scrubbed in to check my work and helped me finish the case. At the end, he patted me on the back and thanked me for "saving the guy's life." Although the gesture and words were simple, they were enough to propel my confidence and complete a stage in my evolution. Going forward, I knew that I could remove a blood clot, rescue a brain stem, and remain calm while doing so. At that moment, I felt a little more like a real neurosurgeon, and a little less like the poseur I had worried that I was.

In retrospect, the line about "saving the guy's life" implied individual credit where team credit was due. I was only one person in a large team that saved his life, including: the stranger who called 911 at the side of the highway, the paramedics, the trauma team, the CT technician, the anesthesiologists, and the OR nurses. Efficiency saved his life. If he had arrived in our ER an hour later, or was delayed in getting up to the OR, I would have been next to worthless.

As with many exhilarating experiences during residency, this one was tempered by the slow onslaught that followed. His recovery was prolonged. He didn't regain full consciousness for days. We watched a steady stream of despondent relatives and friends file into and out of his ICU room. We checked labs, wrote notes, consulted other specialists, performed countless exams, and ordered follow-up scans. We worked to get him off the ventilator. We held family conferences. The questions from loved ones were thoughtful, understanding, and unremitting.

Eventually, he woke up in a confused and somewhat stuporous state. We tried to convey to his family that we were actually pleased with his outcome so far. Recovery takes time. We reminded them how close to death he had been when he first hit the ER. We reminded them that survival was the first goal, and that that was achieved. He should continue to improve. He went off to rehabilitation.

I saw him in the clinic a few months later and he was back to being a walking and talking eighteen-year-old (a "walkie-talkie" as we like to call patients with this level of functioning). Young brains bounce back well. He wasn't normal, though, and still had significant memory deficits. His mother shook my hand and turned to her son: "Remember this nice doctor? She operated on you." He didn't remember, of course. He had no idea who I was. His memory deficit didn't seem to faze him, though. His main concern was whether or not we would allow him to go on roller coasters. He delivered the question in an apathetic monotone drone. (Was that the teenager in him or the head injury? Hard to tell.) Still, I was encouraged. He should look great in another six months or so.

I examined him with my attending, the trauma guru. He pointed out that the incision had healed nicely but that next time, I shouldn't take it quite so low along the front of the ear. I made a mental note and acknowledged to myself that I was still a peon in this lengthy training process. Then, glancing into the patient's equal pupils, I told him to always wear his seat belt.

——

As an aside, trauma patients like this are great examples of the enigma of consciousness. After a severe head injury, severe enough to cause coma (a prolonged absence of awareness), the recovery process eventually reaches a fork in the road. A patient will emerge from the coma—which is a temporary state—to either regain consciousness or enter a persistent vegetative state. In other words, he may wake up or he may not.[1]

The term "wake up" may be confusing, though, as a person in a persistent vegetative state actually has sleep/wake cycles, with eyes open during the wake cycle (this is part of what distinguishes it from coma). The patient is not truly "awake," as in aware. The opened eyes, sometimes wandering, can be misleading and mistaken for conscious interaction, as they were in the famous Terry Schiavo case, where the parents battled the husband over the feeding tube, claiming it should remain in place. It was pulled, and on autopsy of the brain, she was found to have no anatomical capacity for vision.

In the case of a head injury victim, we may round in the ICU at six a.m. and find that the patient responds, at best, in only a reflexive manner to various stimuli. Then, upon rounding again at six p.m., we may be delighted to find that he has crossed over the threshold into a conscious state, opening his eyes and showing us two fingers when we ask. What happened to his brain over those intervening twelve hours? The short answer is: no one really knows.

The serious scientific study of consciousness has been in vogue only relatively recently. It was previously considered too messy a topic, better suited to philosophers and theologians than to scientists. Now, the search for the "neural correlates of consciousness" is on. How does the physical substrate of the brain—individual neurons, chemical interactions, large-scale neural networks, the brain as a whole—produce awareness, self-reflexive thought, the appreciation of existence? There is no shortage of debate. Some scientists believe that consciousness is a sort of "emergent" property of the entire brain's functioning, not specifically a physical phenomenon, but more an elusive property that cannot be pinpointed or studied directly.

Dr. Francis Crick, of the famous Watson and Crick who discovered DNA in the early 1950s, strongly disagreed. He spent the last several years of his life in a quest to answer just that question, before his death in 2004. In his work with Dr. Christof Koch of the California Institute of Technology, he came to the conclusion that consciousness is most likely mediated by relatively few specific neurons (maybe thousands, rather than millions or billions) at any given time.[2] He even speculated that a thin ribbon of brain tissue, known as the claustrum, located on both sides of the brain near the inner aspect of the temporal lobes, may play a major role.[3] It's a reasonable guess, given that this area of the brain has been found to be connected to just about every other region—connectivity must be key—and that its functioning is otherwise not understood at the present. But it's not much more than a very well-educated guess.

On a related note, Koch wonders whether a newborn baby is truly conscious from day one.[4] Although this question may remain permanently in the philosophical realm, it's ripe for quirky dinner conversation with open-minded friends, if not for scientific study. I could see how consciousness might arise over time as the brain develops during the ensuing days, weeks, or months and forms critical connections not yet present at birth, similar in a tangential way to how the brain of a comatose patient flips on, mysteriously, after an incubation period.

Although neurosurgeons witness this mysterious flip of the switch in their own patients, only a rare few have been directly involved in consciousness research per se, and articles about consciousness are rare in the neurosurgery journals. One exception is an article written by the aforementioned Dr. Crick, published in the journal *Neurosurgery* after he was invited to give a lecture at one of our national meetings. The last paragraph reads:

"Consciousness remains one of the major unsolved scientific problems of this new century. The solution of it may well change our whole view of ourselves. We hope very much that neurosurgeons, with their privileged access to the human brain, will join in the search for the (neural correlates of consciousness) in one way or another."[5]

I can speculate as to why so few neurosurgeons have broached the subject in depth. Although the mystery of consciousness is a fascinating one, maybe even *the* most fascinating, the more practical concerns tend to be most pressing, like the patient who is rolling through the ER doors, or having a seizure in the ICU, or bleeding on the OR table.

———

Another punctuation in my evolution as a junior neurosurgery resident involved being stuck in the OR, by myself, in the face of profuse and ongoing bleeding. This one requires a brief anatomy lesson. Some of the most high-volume bleeding that can be encountered in brain surgery arises from what are called the dural sinuses. For that reason, we try to stay out of them whenever possible. These sinuses are distinct from the better-known sinuses underneath the face that can cause headaches or sinusitis. Those sinuses are totally different.

They are filled with air. (We try to stay out of those during surgery, too, but for a different reason: infection.) The sinuses that I'm talking about are filled with blood. This blood is of the darker venous variety as opposed to the brighter, more highly oxygenated arterial variety, because the dural sinuses are charged with the task of draining used blood from the brain.

There is a complex network of sinuses, and they all come together at the base of the skull, ultimately draining into the jugular vein. Most people have heard of the "jugular," and the word usually conjures up images of cheetahs stalking gazelles and "going for the jugular." (In an effort to stamp out multisyllabic words, or maybe just to be cute, surgeons sometimes refer to the vein simply as the "jug.")

I was at a dinner party recently and introduced myself to another guest. When he learned that I was a neurosurgeon, his curiosity perked up, not because he had any particular interest in neurosurgery, but because he knew I would be qualified to settle a bet he had made with a friend. He had seen the movie *Hannibal*, in which Anthony Hopkins plays the highly disturbed and cunning lead character. Near the end of the movie, Hannibal stages an elaborate and grotesque dinner party in which he plans to feed a guest pieces of his own brain (the guest's brain, that is). In order to do so, he drugs the guy, uses a large electrical autopsy saw to remove the top of his skull (presumably in one large piece), covers his exposed brain with a baseball cap, and sits him down at the head of the table, still groggy from the sedatives. I was a bit embarrassed to say that I had seen the movie, too, so I was able to recall the entire scene, in detail.

The bet centered around the technicalities of skull removal. Was it really possible to remove such a large area of skull without having the guy bleed to death? After all, the procedure was performed in a kitchen, without proper equipment, and with no assistant. I thought the debate was fairly sophisticated as it seemed to presume an understanding of the dural sinuses and the pitfalls of working around them.

My answer was that it would have been nearly impossible to re-move the top of the skull in a living being using only such a crude saw. The largest sinus, called the superior sagittal sinus, runs like a racing stripe along the middle of the head, from the forehead all the way to the back. It lies just underneath the skull, encased by the dura and embedded between the two halves of the brain. It is usually par-tially stuck to the inner skull surface. During surgery, if we have to cross over the sinus in removing a portion of skull, we are very care-ful about freeing it up from the underside, or else it is easily torn open upon lifting off the bone.

If the sinus is torn, the bleeding can be difficult to stop. The sinus cannot simply be tied off as might be done with a vein of lesser importance. (Disrupting normal flow through this sinus can cause a large stroke known as a venous infarct.) Holding firm pres-sure, another common surgical technique used elsewhere in the body, is also problematic. First of all, you can't really hold firm pres-sure against the soft brain. Second, you wouldn't want to hold pres-sure too firmly because the sinus could clot off, which, as I just mentioned, could cause a stroke. Suffice it to say that the Hannibal maneuver would have been neurologically devastating, or fatal, for the dinner guest.

I smiled at my new acquaintance and concluded that Hollywood used artistic license in that scene. Smiling and self-satisfied, he pre-sumably looked forward to a payout and an "I told you so."

Although we try our best to stay out of the dural sinuses, we are not always perfect. A particular procedure that I performed countless times during my training requires creating an opening in the skull just behind the ear. The residents, especially the junior residents, were sometimes referred to as "can openers" because we had to open and close the cases by ourselves, over and over again in almost an assembly-line fashion. After completing the opening, we would call the attending neurosurgeon in for the critical (and most enjoyable) part of the procedure. Often, three cases would be going on simultaneously in three different ORs, with the attending flitting from one room to the next as cans were opened and closed.

The tricky part of the bone work is this: The landmark for the opening is at the junction of two sinuses that come together just behind the ear, the transverse and the sigmoid sinuses. If you don't come right up to this corner, the exposure is inadequate and you will feel like a fool when the attending walks in and realizes that he can't see what he needs to see. Often, a sinus can be recognized by the bluish appearance of the overlying dura or the presence of a small vein entering into it. These rules of thumb, though, are not always consistent or reliable. So, if you're working away at the bone, little by little, and don't recognize that the sinus is just underneath, you could be in for a surprise if you create a nick in the dura with a bone spicule or an instrument. This is quite easy to do, especially in a patient with "old lady dura," which tends to be flimsier than most, and also more adherent to the bone.

I had seen minor holes in a sinus before and was surprised at the volume of blood that could escape. With two pairs of hands working at it, though, the bleeding can be stopped with gentle pressure, a special foam material, and patience. If the bleeding is profuse, we may have the anesthesiologist elevate the head of the bed to decrease the overall pressure on the system (this deceptively simple maneuver can actually help).

So there I was as a junior "can opener" alone in the room. After making the small scalp incision behind the ear, I chose the best area to touch down on the skull with the perforator drill. The drill stopped, as it should, upon completing its full thickness excursion. As I pulled the drill out, though, I was met with a torrent of blood that filled the operative field, started to spill over onto the sterile drapes, and just kept coming. Without a doubt, the drill had torn open the sinus, and the hole wasn't small.

Again, there is no role for thinking here. It had been drilled into my head by the more senior residents that I ask for two things in this situation, automatically: a large piece of gel foam and a large cottonoid. While an assistant keeps a suction device in the field to siphon off the ongoing bleeding, you try to plug the hole with a large piece of spongy material (gel foam) and then cover that with a large cottonoid, which looks like a strip of felt. I placed these items over the hole but still, for the first several seconds, I was certain the bleeding would never stop. There was just too much of it. That thought wasn't helpful, though, so I kept working at it.

It took me a minute or so, but I got the bleeding under control and called for assistance in completing the rest of the opening, which was more laborious than usual. We had to avoid disturbing the rather tenuous floodgates where the hole had been patched. All told, the patient probably lost about 200 cc of blood (eight-tenths of a cup) before I stopped the bleeding. Although this is a relatively small amount, it occurred over a very short time period and was enough to raise the heart rate (my heart rate, that is).

Total human blood volume is about 5 liters, so 200 cc of blood loss would make up only about 4 percent of the total blood volume. A heart-lung transplant surgeon would laugh at this. So would many orthopedic surgeons. Neurosurgeons would argue, though, that blood coming out of the head is distinctly different from blood coming out of the leg.

Despite the tear in the sinus, the patient did beautifully. She didn't need a blood transfusion and was home in three days. The torn sinus was not technically a complication, but I did have to blame my technique, even though I knew it could happen to even the most senior and experienced of surgeons.

I attribute the punctuation in evolution here not only to my control of the bleeding, but also to the response of my fellow residents. After the case, I removed my bloody shoe covers and walked into the residents' office. A few of the senior residents were lounging around and asked me how my case went. "I got into the sinus. The drill just ripped it open." I must have looked dejected.

They all smiled and one piped up: "Did you have your hip waders on?" Riotous laughter followed. They had all been there before. They weren't making fun of me, but instead were inviting me into the club. There is solidarity in adversity, and the fact that I handled it just fine made me one of the guys.

EIGHT

Tools

Neurosurgeons are practically worthless without their tools. In order for neurosurgeons to do their job well, these tools need to stay organized. Almost nothing irks a neurosurgeon more than when a surgical instrument can't be found or handed over quickly enough. This can happen because (1) it is buried in clutter, or (2) the scrub nurse is new. I especially pity the new and disorganized scrub nurse. This fatal combination can bring out the worst in a surgeon.

This became obvious to me when I scrubbed in with a slick senior neurosurgeon known for his efficiency in the OR. We were doing a straightforward spine case and he had a tee time scheduled to follow. He was not in the habit of being late for a tee time. The nurse was new. The word was that she was a former exotic dancer, which, believe it or not, is actually marginally relevant to the story.

About one minute into the case, I could tell that he was getting

antsy. There was too much delay between him speaking the word
"Cobb" and the instrument hitting the palm of his hand. He
couldn't get into his usual rhythm. He glanced up at the clock on the
wall and said, "Let's stop right here for a second." I knew this was
trouble.

He looked Ms. Former Exotic Dancer in the eye and then, in the
softest of tones, explained: "Pretend this is your makeup counter.
You would have things organized, right? Your mascara goes here, lip-
stick there, eye shadow over there." He pointed to different areas of
the instrument table.

It was strange to hear a surgeon like this say a word like "mascara."

"So, we'll put our kerrisons here, Leksells here, and periosteals
there." He arranged the instruments for her in neat little rows. "Now
let's get going."

All concerns of sexism aside, I was happy to see that this little tu-
torial actually worked. The case went smoothly from that point on,
and the surgeon had plenty of time to get to the golf course. Not only
that, the nurse seemed genuinely grateful for the advice. (I knew
that the surgeon intended no offense. I had already judged him to be
a genius in speaking to people on their own terms and in their own
terms. I accompanied him once to see a patient who was a librarian.
He greeted her with: "So, what's new with the Dewey decimal system, huh?" Maybe a little old-fashioned, but you had to love him.)

Because neurosurgeons spend a large portion of their professional
lives in the operating room, it's not surprising that we can develop a
certain affection for the tools we use. Although I admit to having my
own favorites (I like the simplicity and versatility of the slim, no-
frills Penfield #4 dissector), some neurosurgeons go a bit overboard,
assigning cutesy nicknames to their favorites, names that seem a bit
out of character in the macho world of surgery.

One attending I worked with, in the middle of performing the mechanical grunt work on the skull necessary for invasive skull base

surgery, would ask the scrub nurse for "my little nipper." This is not the instrument's given name. It is his, and only his, nickname for a diminutive variant of a tool properly called a "rongeur" (the *g* here being the French type), designed to bite off pieces of bone. New nurses are expected to know, a priori, what "my little nipper" is, and some actually figure it out, without prompting, by scanning the table of possibilities and reading the surgeon's mind. Others don't even try, raising their eyebrows and asking: "You want the *what?*" After swallowing my pride, I gathered up the courage to ask for "my little nipper" when working with this neurosurgeon, assuming that consistency within the team was important.

Claiming possession of an instrument that you don't own or that you didn't invent can rub people the wrong way. One neurosurgeon I know made frequent use of a dissecting instrument called a "freer," which is pronounced in two syllables, as in to free something up. Case after case, use after use, he started to ask for "my freer, please" rather than just "freer" or "the freer, please." In the setting of a long, multihour case, this subtle shift in nomenclature annoyed one of my fellow residents so much that, when that neurosurgeon stepped out of the room during an operation, the resident turned to the scrub nurse and said, in a mock-serious tone, "I'll take the 'myfreer' please," as if it had become one word. Long hours in the OR have a special way of magnifying trivial annoyances.

Other nicknames are just cute abbreviations, like "pitute" for "pituitary rongeur," and some instruments have names that are already so cute they don't need a nickname, like the "peapod" instrument we use in disc surgery. Certain well-established nicknames make no sense, like the "bunnies" whose proper name is "Adson forceps." (I refuse to ask for "bunnies.") When it comes to requesting an instrument to retract against the brain, I have my choice of three different names (brain retractor, malleable retractor, and brain ribbon) for the same instrument. I prefer "brain ribbon" because it is, by far, the most poetic.

Although I've never claimed an instrument as my own and I've never coined any affectionate nicknames, I do have a soft spot for surgical instruments in general, partly because of my upbringing and partly because I like design, and so many instruments are a perfect coupling of form and function that I'd say are worthy of display in the Museum of Modern Art.

I was in an antique store recently with my husband and he pointed out an interesting artifact in a glass case that he knew I'd love. It was an old hand drill used to make holes in the skull, long before the era of the slick power drills we use today. As I'd seen with other antique instruments (and been amused by), there were even a couple ornamental elements, a touch of design beyond what was required for function, almost as if it were meant for display.

My first impulse was to buy it. But then, my husband and I started thinking: What would we actually do with it? Would we display it in our home as a morbid conversation piece? Could I display it in my office, or would that be completely inappropriate, prompting nervous questions by already nervous patients? Our indecision, combined with the price, made me pass on the opportunity, and I kind of regret it now, although I still don't know what I'd do with it.

A small hospital that my father used to work at closed several years ago, and the OR was throwing out tons of outdated and broken equipment. They must have felt the stuff wouldn't even have been worthy of donation to a third world country. My father—a natural tinkerer with an inventive mind—picked through some of the "rubbish" and brought several selected pieces home with him. For months, you could come across an old used endoscope here and there, in the family room or the library, and you might wonder what orifice it had last been pushed through.

After a while, the novelty of using the scopes to look under bookcases wore off, and our family found no further use for them, until my brother came home one day with a brilliant idea: eBay. He had become accustomed to foraging through my mother's long-forgotten basement treasures, Hummel figurines, and once-precious china, selling the goods on eBay to help pay his Brooklyn rent. The endoscopes then, transferred from the hands of a tinkerer to the hands of an opportunist, made their way into cyberspace and were quite a hit. They sold like hotcakes. (Who bought them, I wondered, and for what potentially subversive reasons?)

I guess it's okay to profit off of someone else's trash, but it does give me pause. I have to think of it this way: it would otherwise have ended up in a landfill. But do you owe anything to the person or organization that was throwing the "trash" away? My mother would answer with this famous story of hers. She was leaving a friend's home, driving down the driveway, when she noticed an old worn-out oriental rug lying on top of a garbage pail, awaiting pickup by the garbage truck. She went back and asked the friend if she could take it. Her friend, of course, had no problem with her taking it—it was trash.

My mom used the rug in our garage for a while, but my father tired of it and requested that she get rid of it. My mom cleaned it up and donated it to a local philanthropic organization that automatically had it formally appraised—at a value of $1,000. A while passed before she decided to tell her friend, but they both had a good laugh, and watched what they threw out from then on.

———

Although the variety of tools that we have to choose from is enormous, we tend to use certain ones over and over again, from case to case. This does cut down on the confusion a little. A favorite saying of one of my mentors—"Ah, the sound of neurosurgery!"—refers to the sound made by the most commonly used instrument in modern brain and spine surgery, an instrument we deem absolutely essential. We demand two for every case: one for the main surgeon, one for the assistant. We rely on this instrument throughout the entire operation and use it to perform the most basic and the most complex of functions.

Granted, neurosurgery is a technically advanced field, but in this case I'm not referring to lasers, robotic assistance, or holographic imagery. I'm talking about the lowly suction device: a thin metal suction tip connected to long clear plastic tubing that's hooked to a centralized wall-based vacuum system. This instrument is nearly identical to the one used by dental hygienists to clear the saliva out of your mouth during a dental cleaning, except that their suction tips are usually plastic and disposable. We reuse ours.

During an operation, the surgeon's nondominant hand is nearly always occupied by holding and manipulating the suction, while the dominant hand controls other instruments, such as the drill, the bipolar (a coagulating device), scissors, or a dissecting tool (of which there are over a dozen to choose from). We tend to call this instrument the "suction" or "suction tip" rather than "sucker."

The suction serves two main purposes: to retract against various tissues, including brain, and to continuously clear the surgical field of fluids that get in the way, namely blood and cerebrospinal fluid. (Hence the omnipresent, somewhat annoying, sound of fluid being sucked through the system, the "sound of neurosurgery.") The suction tip can also be used to help push things into place, such as when packing various materials into a hole to stop bleeding. The suction, though, has more than just a supporting role. When removing a soft brain tumor, the bulk of the work may be accomplished by the lowly suction, even in this modern era.

Other surgeons have been known to make fun of us for our strongly suction-centric ways. (All surgeons use a suction device, but not quite so extensively.) While general surgeons, cardiothoracic surgeons, and orthopedic surgeons toil over abdomens, chests, and limbs, you will notice more cutting, sewing, and tying. In short, they look like they're doing surgery. Their hand movements are grander. They may even have to get their elbows and shoulders involved for the bigger moves.

Neurosurgeons doing brain surgery, on the other hand, tend to look like they are picking at things, sucking things out, little by little, sometimes for many lonely hours at a time. Once the grunt work of bone removal is done and the brain is exposed, the microscope is wheeled into position. The moves are small and delicate. Elbows and shoulders remain still.

Despite, or maybe because of, their simple tubular structure, suction devices are not foolproof. They are a leading source of mundane frustration for the neurosurgeon in the OR. Pieces of tissue or clotted blood often clog the suction tip or the tubing, requiring the surgeon to interrupt the flow of the case, hand the device over to the nurse, and have it flushed out with saline irrigation. When that fails to clear the problem, all eyes turn to the large canisters by the wall. Are they full? Or, perhaps the quiet and unsuspecting medical student standing in the background is the culprit: Is his foot on the tubing that runs along the floor? If so, he is forgiven, but only once. He'd better not step on it a second time, or the surgeon may start to question his worth as a human being.

If the suction is our number-one most basic instrument, the drill may be number two. That's how we get into the head. Obviously, drilling holes in the head is one thing that distinguishes neurosurgery from the typical desk job. Not so obvious, though, is that the drilling of a hole can be more meaningful than you think. The honor of placing the first "bur hole" of an operation is a gift, of sorts, that we sometimes grant to our eager interns who plan to give themselves over to neurosurgery. We present them with a freshly exposed portion of skull, scalp retracted off to the side, and then ceremonially hand over the weapon. The honor of having drilled into a live human skull then affords the intern bragging rights: "Yeah, I placed a bur hole today. The guy's fine." The intern may say something like this, nonchalantly, to his colleagues who are bored stiff from checking labs and writing orders all day.

As one of my mentors is fond of saying: "Surgery is controlled trauma." When you see the device we use for drilling holes, this statement becomes clear. We have a special drill, a "perforator," that is designed to make this hole-making task foolproof. The drill bit is complex in its geometry and formidable in size. Its function, though, is quite elegant. You have to apply firm and constant downward pressure during the drilling process, but the cutting action stops, miraculously, as soon as you're through the bone, thereby protecting all that is soft underneath. (That's why we allow the lowly intern to get a piece of the action—it's really the drill that we trust.) What you are left with is a nice smooth, round, full-thickness hole in the skull, roughly nickel-sized in diameter. Believe me—this was a great breakthrough in the history of neurosurgery.

For all its wonders, though, this drill does have at least one downside. An annoying glitch in the technology makes it nearly impossible to continue using the perforator on the same hole if you stop drilling halfway through. It won't reengage. Quite disheartening, this is yet another thing that can interrupt the flow of a case, necessitating either chipping away at the remaining bone with a sharp scooping instrument or setting up a different type of drill, with a smaller head, to complete the task. With this glitch in mind, whoever is teaching the intern to use the perforator is tempted to yell "Don't stop!" above the loud drone of the drill. I have seen, on more than one occasion, the driller stop drilling in order to clarify what is being yelled to him.

A favorite teaching point, passed perennially from professor to resident to intern, involves listening to the subtle differences in sound that the drill makes as it passes through the different layers, or "tables," of the skull. It requires only a couple bur holes' worth of experience before you can detect, by sound alone, when the drill is about to stop. This is not rocket science, but it is neat. Again, such knowledge distinguishes us from the guy on the street or at his desk in the office park.

I can't talk about drilling holes in the skull without mentioning one of my favorite neurosurgical terms: bone dust. You may be wondering (or maybe not) what happens to the core of bone being drilled through. Basically, it piles up along the margins of the hole. However, because the assistant in the case usually drips saline irrigation over the drill bit to prevent it from getting too hot, the bone dust often turns into more of a hydrated bone meal.

Early in my training, I was gently scolded by one neurosurgeon for irrigating too vigorously and washing all the bone away. I didn't realize that his custom was to collect the seemingly useless bone dust and use it to fill in the holes at the end of the case (theorizing that the hole would seal over faster and more completely). I had learned from others, instead, to cover the defects with thin, round titanium plates—a more contemporary approach. As a trainee, it became clear to me that I needed to learn not just the science and the mechanics, but also the gamut of personal preferences. The best senior residents remember to forewarn their juniors before unleashing them into various operating rooms with particular neurosurgeons: "Don't forget. This guy saves the bone dust." Such helpful reminders can prevent the needless reprimanding of one adult by another.

Bone dust has a certain smell. It's fairly subtle and not really offensive, but you know it when you smell it, especially when the assistant is lax in dripping the saline and the dust remains fine and dry and flies around the room. As a neurosurgery resident, I often returned home late after a day confined to the operating room and had to sneak into bed next to my husband, who was also a neurosurgery resident at the time. If he awoke fully enough and his olfactory skills were sufficiently tuned, he would complain: "You smell like bone dust." If he fell back asleep right away, I'd be fine. But if he persisted in his complaints, then I was obligated to rinse off in the shower before getting back in bed. Luckily, I always reserved equal rights to complain if he was the one slipping in late, so it never became a source of inequality in our marriage.

In most brain operations, at least a few holes are created and then connected with a different type of drill—one with a foot plate. In this way, an entire "bone flap," or section of skull, can be removed. This part is usually done fairly rapidly so as to get it out of the way. Once the brain is exposed, things take a turn for the more serious and the pace may slow down considerably. In fact, if the surgeons are listening to raucous "opening music"—usually the younger ones— they may request that the music be changed or turned off. When the hard part is over and it's time to put the bone flap back on, some will request "closing music," which is understood by all in the room to be of an even rowdier variety.

I don't want to leave the impression that neurosurgery is all just low-tech drilling and sucking. From my experience in speaking to guests who visit our operating rooms, the biggest crowd pleaser is one of the more complex devices we use: the 3-D image-guidance system. It's the kind of cool technology that a layperson expects us to use. There are many competing varieties, with brand names meant to appeal to the techie in every surgeon, like Stealth and BrainLab. These systems, now commonplace, were not widely available several years ago. We now rely on them to avoid the embarrassment faced in years past by our older colleagues: rooting around trying to find a tumor buried deep within the brain, or unwittingly leaving a large portion of it behind.

Although this may be hard to believe, the following exercise can be a challenge for a neurosurgeon: have a patient lie in front of you, hang the patient's MRI up on the light box, and point, on the patient's head, precisely to where the tumor should be located. It's easy to be off by an inch or even two. Why is this exercise difficult? Mainly because the head is round.

Before image-guidance systems were available, neurosurgeons would err on the side of using a large incision and creating a generous window into the skull. By maximizing the playing field, they improved their chances of being able to find the tumor. In the modern era, with 3-D image-guidance, the patient's MRI scan is downloaded into a computer system in the OR, and these images are linked to a navigation wand. When the surgeon touches the wand to the patient's head at the time of surgery, a pointer appears on the MRI, indicating the precise location of the wand on the head. This allows us to have a kind of X-ray vision. With the guidance of this magic wand, the incision and the bone flap can be centered directly over the tumor, allowing for the smallest possible opening.

Once the neurosurgeon starts to remove the tumor, the wand can be used again to point inside the brain, within the tumor, in order to gauge how far he or she has gone. This is helpful in preventing a surgeon from being either too aggressive (going beyond the borders) or not aggressive enough (leaving too much behind). With certain types of tumors, it can be difficult to tell exactly where the tumor ends and the brain begins. The MRI may present these margins more distinctly.

I can remember assisting on one of my first brain tumor cases as a junior resident. Once we got into the tumor, my mentor tried to verbalize the subtle differences between tumor and brain, in appearance and consistency, as he worked and I watched. I just didn't see it. I took his word for it but remained a bit skeptical. Later in the case, when he granted me the opportunity to use the metal suction tip as an extension of my own hand, I understood what he was talking about. Still, it can be a little unclear at times, even in experienced hands.

A few Luddites in our department pooh-poohed this image-guidance technology at first. They worried that the young ones would come to rely on it so much that we wouldn't think for ourselves and wouldn't bother to cultivate the innate 3-D capabilities of our own minds. Their fear may have some truth to it, but patients certainly don't mind us taking advantage of this technology. The Luddites' perspective would be similar to preferring a bank teller to add all your deposits with a pencil and paper. While certainly possible, it's always reassuring to see them use a calculator.

Sometimes our desire to have all the latest technology can go a little overboard. When our department found out about a new surgical microscope that incorporated an advanced laser sighting mechanism, we had to have one. It sounded incredible. A special room in the OR was outfitted so that this ultra high-tech (expensive) piece of equipment could be mounted to the ceiling and could descend, electronically, into position when needed. Two of our residents were flown to Europe to get the special training required to use it.

With the OR renovation complete and this formidable multi-limbed, multijointed microscope mounted and ready, the rest of the department went through brief training sessions. In these simulated exercises, the device seemed to work well enough, but there were the small glitches you might expect from anything new, complex, and mechanical.

I remember sitting in the surgeons' lounge between operations and hearing an onlooker's account of a case that had just been done with the new microscope. "They might as well have been looking through a cardboard tube!" Apparently, the device didn't quite live up to everyone's expectations. Soon after the fanfare of its introduction died down, it tended to remain suspended high up against the ceiling, in hibernation. Case after case, one of our traditional operating microscopes would be wheeled into position underneath it, and after a while, no one even seemed to look up at it anymore. It wasn't all for naught, though, as it remained an impressive fixture that we could point out to professors and fellows visiting from other, less technically endowed, programs.

———

Given my sometimes excessive introspection into my own career, I often reflect on the fact that the experience of performing surgery can span the spectrum from pure enjoyment to tedious boredom or even frustration. The most enjoyable cases (for me, at least) involve the greatest number of the following elements: a pleasant, familiar OR team, a "fresh" case (as opposed to a scarred-in "redo"), sharp and well-maintained instruments, a thin patient, glitch-free technology, good music, performance of the case during normal working hours, and a quiet pager. Less enjoyable cases often involve opposite elements. One quickly realizes as a surgeon that only certain elements of the perfect case can be controlled. Given the fact that most surgeons can be classified as control freaks, it can be hard to accept this reality.

For me, I am at risk for boredom when things become overly routine or if I'm not learning anything new, which is a problem, because in surgery, you should strive for routine and a consistent focus. Patients aren't really looking for a creative surgeon or even an intellectually curious one. The ideal surgeon would be one who doesn't mind doing the same cases, over and over again, the same way, with the same instruments, year after year, continually enhancing safety and efficiency while building case volume. In such a scenario, although the mind may be at risk of going a little numb, the hands may continue to enjoy going through all the motions that they know and do so well.

I can't ignore the most obvious benefit of the well-oiled surgical routine: you get to help people. This certainly makes up for some of the downside of routine and can be quite satisfying, emotionally, unless the fear of lawsuits and the compromise of family life cancel out some of this benefit.

The profession of surgery, then, presents an interesting state of affairs: it attracts some of the best and brightest, but is heavily reliant on tools, manual labor, routine, sacrifice, a large support staff, and a narrow focus. This combination can represent, at extremes, the perfect sweet spot or the perfect recipe for burnout.

NINE

Risk

Hooking a fish at the edge of its mouth doesn't really bother me, even though you have to slide the hook out backward, tugging the pointed barb through the skin, in order to extract it. It's a necessary part of fishing. I watched my dad do it when I was a kid, and he assured me that the puncture site heals just fine. I would watch the fish swim away after he threw them back in. The small flesh wound didn't seem to slow them down at all.

Every once in a while, though, a fish will swallow the hook. This is not a simple flesh wound. If the hook makes it into the stomach, it can emerge from the mouth with shiny, dark red viscera trailing behind. Sometimes, the hook can't be extracted, so the line is cut, leaving the silvery relic inside. I've watched these fish swim away, too, but I knew they were doomed. For me, the swallowed hook tarnishes the otherwise carefree game of catch-and-release, at least for a little while.

One of the patients who had the most enduring impact on me was one I knew the least. I didn't even get to meet her before surgery. She wasn't my patient. I was just the fourth-year medical student rotating on the neurosurgery service, excited to participate in a cool case. The case was cool because it was big and complex. Risky. Despite my enthusiasm, I didn't delude myself. "Participate" was too strong a word. At my level, I would be relegated to scrubbing in and watching. The chief resident made me feel like part of the team, though, by discussing the case with me and granting me the dubious honor of placing the Foley catheter in the patient's bladder, a lowly but necessary task. I also took the initiative to write some orders in the chart based on what I knew she would need after surgery. These orders would turn out to be unnecessary.

I learned from my chief resident that the patient, intubated and asleep in front of me, was a young woman, a teenager really, who decided to undergo surgery only after painful deliberation. Years earlier, she had been diagnosed with a large malformed tangle of blood vessels in her brain (an "AVM," or arteriovenous malformation). Unfortunately, this AVM was of an extreme type—very large and in a very dangerous location—informally known among neurosurgeons as a "handshake AVM." As you walk out of the neurosurgeon's office, a handshake is all he has to offer.

So, for years, the patient and her parents lived in fear, never knowing if or when this malformation would decide to bleed. They knew that a bleed could be fatal. They also knew that surgery could be fatal. They respected their surgeon's seasoned opinion that surgery wasn't an option for her. They understood his reluctance to risk having his own hand in her death or, worse, her neurological devastation if surgical removal were attempted.

One surgeon's handshake became another surgeon's challenge. When her original pediatric neurosurgeon left town to practice elsewhere, she and her parents sought the advice of another neurosurgeon, one known for both his superlative microsurgical skills and his willingness to take on the most difficult cases. It was unusual for him to turn away a case. In fact, on one rare occasion, in advising a patient against surgery, he was rumored to have told her: "You don't need me. You need Jesus Christ."

This surgeon's way with words was as well known as his surgical skill. In making a teaching point to a resident examining a patient in his clinic once, he wanted to explain how the patient's disorder was not of the familial genetic variety passed from one generation to the next but was, rather, the product of a random spontaneous genetic mutation. He pointed to the patient and explained: "She's a spontaneous mutant."

I suspect the young woman and her parents were impressed by this surgeon's confidence and reputation. Their impression, combined with the chronic unease that arose from doing nothing, must have tipped their decision toward surgery. In essence, a decision like this comes down to: Do you want to take your risk up front, all at once (surgery), or slowly, over time (wait and watch)? Individual personality, more than science, can be the driving factor in making such a choice.

The operation was a challenge, a technical tour de force. The AVM, which had probably been there since birth, did not give in easily. It had spent its entire existence within the dark confines of her skull, sharing space with her brain, and her brain had unwittingly accommodated its presence. Although a potential threat to her life, the malformation was a native and natural part of her, not a recent invader.

The surgeon worked for hours, meticulously, under the bright focus of the surgical microscope. He closed off one abnormal blood vessel after another, making sure to interrupt the complex inflow to the beast first, knowing that interrupting its outflow too early could provoke a bloody explosion.

The final vessels were closed off and the tangled mass removed. Her brain now remained as the sole inhabitant of her skull. I was surprised by the size of the depression left behind, where her brain had accommodated to the malformation's presence. Her head was closed up and she was wheeled out to recovery.

After witnessing this surgeon's skill with my own eyes, I agreed that his reputation, and even his cockiness, was well deserved. If I needed brain surgery, he would be my surgeon. I thought about how satisfying it must be for him to go out to the family, announce his success, and vindicate their most difficult decision. They put their daughter's life in his hands, and he was able to offer her a life without fear of the malformation that had been so intimate with her brain. Others had warned strongly against surgery, citing unacceptable risk. They went ahead anyway, and could now be grateful that they had made the right decision.

The patient woke up gradually over the next half hour, recovering slowly after hours of anesthesia. She wasn't awake for long, though, before the nurse noticed early signs of trouble in her neurological examination. Minutes later, she was unresponsive. A stat head scan revealed a catastrophe: massive bleeding into the brain, including the delicate brain stem. The surgeon went through all the right motions of a heroic rush back to the operating room, but the damage had been done and he knew it. The bleed was fatal.

Despite all good intentions and a technically successful operation, her brain could not tolerate the perturbations in circulation that accompanied removal of the large tangled mass of vessels. Maybe an otherwise normal artery in her brain, not used to the new pressure dynamics, broke open. Or, a critical vein near the malformation may have clotted off, leaving too few outflow options for the brain's rich blood supply. Whatever the explanation, I imagined that this was the AVM's final demand for respect, with her scan representing a "Don't Touch" warning to other surgeons tempted to offer others like her more than just a handshake. It was also a tragic introduction to the mantra I would hear again and again through my training: "The patient is the one taking the risk, not the surgeon."

Years later, as a senior resident, I met another patient with a handshake AVM. She had received only handshakes and had resigned herself to inaction long ago. This woman's AVM was so large that it extended across the corpus callosum, one of the structures that connects the two hemispheres of the brain. Although she was otherwise a healthy and active woman, in her thirties, she had lived her life with full knowledge of the tangled mass that would always be with her, a cohabitant that would demand extra attention every once in a while.

This woman had never suffered a devastating bleed. Instead, there were a few defined episodes in which the malformation leaked fairly small amounts of blood into the brain. (This scenario is actually fairly typical for the largest of AVMs. The smaller ones are more likely to cause larger bleeds for various reasons.) Luckily, these small bleeds were in the relatively resilient frontal lobes, and the patient suffered bad headaches but no significant neurological sequelae. When I met her, she was in the hospital for a few days after one of these bleeds, and my job was to check in on her and make sure her blood pressure and her headaches remained under good control. That's about all we had to offer and, luckily, that's all she needed.

Had these two patients, victims of random developmental circumstance, been given the chance to meet each other, what advice would the elder have given to the younger? It's clear that the brain can accommodate quite nicely to the overbearing presence of a malformation, but can the mind be trained to accommodate just as well? When inaction is the best action, how do you prevent fear itself from becoming an illness? Does the fear simply wear out, or does it have to be forced out?

———

Knowledge is power, but it can also be fear. Surgeons are obligated to educate a patient about their condition and treatment options, but then they are faced with managing the anxiety that goes hand in hand with that knowledge. I have found that handling a patient's anxiety can be more complicated, and sometimes even more time-consuming, than the surgery itself. Some surgeons loathe this part of the job. It reminds them of all the reasons they didn't go into, say, psychiatry. They prefer patients under anesthesia to patients wringing their hands, crying, and reading off a list of questions from everyone in their family, including their second cousin. Others find those interactions rewarding. I tend more toward the latter camp, but I do empathize with those in the former because I understand the surgical personality and have just a touch of it myself.

Because anxiety management is not always enjoyable, some surgeons don't spend much time on it. I remember, as a resident, having to recalibrate a patient's thoughts. She was convinced that she was dying of a brain tumor. She had a small benign tumor, called an acoustic neuroma, on one of the nerves at the base of her brain. She had no symptoms. The tumor was discovered incidentally, when her head was scanned for other reasons. She was elderly, and a surgeon at another institution recommended doing nothing for it. She left his office with this sentiment: "I have a brain tumor and nothing can be done for me."

I saw her and her extended family a few months later, when a relative urged her to seek an opinion at our institution. She looked around at her loved ones in the room and expressed regret that this would probably be the last Christmas she would be spending with them, as death was near. I glanced down at the name on her chart. Was I talking to the right patient?

I went over her MRI and examined her. I explained the reality of her small benign tumor at the base of her brain (not *in* her brain), and told her that it could have been there for quite a while. Most likely, she would die years down the line of a totally unrelated cause, before this little tumor could even cause a significant problem. I went over all the options and we settled on the one everyone was most comfortable with for the time being: observation. I was happy to be of service as it is always gratifying to extend someone's life expectancy without even having to pick up a scalpel.

During my training, I took to observing how different neurosurgeons interacted with their patients in discussing the risks of surgery. I knew I'd have to devise my own personal style, but I figured I could pick up on what seemed to work and what didn't. On one extreme was the warm hand-holder who peppered religious-speak into his counseling, adding blessings to his discussions of what could possibly go wrong. ("We'll get you through this, with God's grace.") I have to be honest. That style did work wonders, especially with the older ladies, but I could never adopt it myself. I wouldn't be able to keep a straight face. The same surgeon was effective in conversation in other, more creative, ways as well. I observed him discussing a difficult situation with a patient and her very large, extended Italian family. He was trying to get across the fact that the tumor at the base of her brain would be tricky to remove because of all the nerves draped across it. After thinking about it for a few seconds, he explained, "It's like trying to get at a large meatball when there are strings of angelhair pasta in the way."

On the other extreme was the guy who, I'm a bit ashamed to admit, was entertaining to watch, in a sadistic sort of way. There's only one word to describe his style: blunt. Here's how he would describe the risks of surgery for an aneurysm of the brain, just prior to having a patient sign their consent: "You could have a stroke. (Pause.) You could have permanent brain damage. (Pause.) You could become a vegetable. (Pause.) You could die." Although these statements were technically correct, the monotone voice with which they were spoken, and the sharklike demeanor that went with them, explained his uncanny ability to make a patient and their family burst into tears.

Needless to say, I didn't adopt this style wholesale, either, but I did appreciate the warning this surgeon left me with: if the patient isn't crying by the time you're done going over the consent for surgery, then you haven't done your job. Although I don't force an upwelling of tears from each and every patient, I agree with the spirit of the advice in that the risks of surgery have to be laid out plain, in the open, and cannot be taken lightly. And even though some patients prefer not to hear all the risks and just want to get the signing over with (worrying that if they hear too much, they'll change their mind), I think it's in their best interest to know everything anyway.

From a surgeon's point of view, the last thing you want is for a patient to come back after surgery saying they had no idea they could end up with: an infection, headaches, nerve damage, a numb foot, an ugly scar, a less than perfect outcome (take your pick). The next person you'll hear from is a lawyer. The growing cynic in me, though, says that nowadays, you may hear from a lawyer even if the patient did fully understand all the risks.

Some surgeons late into their careers grow weary of repeating the same spiel to patients, over and over again, explaining the risks, benefits, and alternatives (the standard triad) of the same procedures. One mentor of mine, a senior, big-picture kind of guy, would leave the consent process to his minions, usually the residents. The story goes that if a patient asked him directly about the risk of stroke in a procedure he specialized in, he would wave his hands while standing up to head for the door and explain: "We did have a stroke once. Years ago. Older woman. In her eighties. Lied to us about her age." I never actually heard him say this myself, but the story was repeated so often and by so many people that I fully believed it to be true.

Some surgeons shun statistics. Others love them. Some patients want exact numbers, others couldn't care less. The conversations, then, between a surgeon and patient can have many different flavors. In one, you could hear precise percentages and the quoting of recent scientific literature. In another, vague words such as "possible" and "unlikely" would dominate the discussion of risk. Most surgeons are not statisticians, though, and so even when detailed numbers are discussed, they may not be perfectly accurate.

Colleagues of mine wrote an interesting paper in the neurosurgi-

cal literature that exposed a common misunderstanding among neurosurgeons.[1] Take this situation: You have a patient with a newly diagnosed AVM of the brain, and it has never bled before. Let's say the
literature estimates a roughly 2 percent chance of AVM rupture per
year, based on large population studies. How do you counsel this patient about management options?

Age, of course, would be of critical importance. If the patient is
ninety years old, you would probably advise the patient to leave well
enough alone. They will probably end up dying of another cause
within the next few years anyway, and interventions designed to fix
the AVM come with their own set of risks. Plus, the AVM may have
been present, lying dormant, for decades. At age ninety, better to just
live with the relatively small risk of doing nothing.

What if the patient is sixty-five years old, though, and asks you
what the chances of rupture would be over the rest of his life if nothing were done? You could estimate a life span of, say, eighty-three
years. He has eighteen to go. Does that mean that the chance of rupture is eighteen times 2 percent, or 36 percent? It turns out, based on
a survey, that around half of the neurosurgeons felt this was the correct way to come up with the answer (the other half were all over the
map). It turns out that it's not.

Look at it this way. What if the patient is only fifteen years old? In
this case, the patient could be estimated to live roughly sixty-two
more years based on life expectancy tables. Does this mean the chance
of bleeding over this patient's life is sixty-two times 2 percent, or 124
percent? That makes no sense. It turns out that an accurate assessment
of lifetime risk is a more complicated, logarithmic-based equation:

$$\text{Risk of hemorrhage} = 1 - (\text{risk of no hemorrhage})^{\text{expected years of remaining life}}$$

The real answer would be 85 percent, not 124 percent. In the case
of a young person, the exact answer doesn't really matter because
any neurosurgeon would tell you that the AVM should be fixed. Re-

gardless, the point is that counseling based on hard numbers is not as straightforward as it may seem.

————

I saw a patient recently who had had spine surgery performed a few years earlier by another surgeon. As can often be the case, the original reason for the surgery—advanced arthritis, or garden-variety "wear and tear" changes that can occur with age—continued to progress and she was now faced with requiring treatment, and maybe even a second operation, for a neighboring level of her spine. I knew the surgeon who performed the first operation, a highly reputable colleague, and I voiced some question as to why she wasn't in his office instead.

"Well, *he* gave me a wound infection, so you can be sure I won't be going back to *him*!" This sort of statement, and the vehement emotion that goes with it, worries me and raises a red flag. It might be easy for me to fall into the trap of flattery, at first (the patient specifically chose me over the other surgeon), but the realistic response should be one of caution. This is the type of patient who believes that the concepts of risk and complication are neatly and inextricably linked to another concept: blame. If something bad happens, it's someone's fault. There is no such thing as bad luck.

Based on the alarmist tone of her voice, I imagined that in *her* mind, the surgeon willfully smeared bacteria into the surgical site, aiming to set up an evil chain reaction leading to fever, pus, and a red, swollen incision. The truth is that infection remains (and will always remain) a risk of any surgical procedure. Although all measures are taken to bring the rate as close to zero as possible, it still hovers around 1 percent or so (or slightly higher or lower, depending on the surgical site, the circumstances, and how healthy the patient is to begin with). Surgeons feel terrible when a patient of theirs develops an infection, but they normally don't feel guilty. While it's true that in very rare cases, careless breaches in sterile technique are to blame, and certain individuals can be held liable, that is the very rare exception.

So, if you are the unlucky individual who happens to fall into that 1 percent because natural bacteria on your skin (the usual source) infected your wound, should you blame your surgeon? Should you call your lawyer? Should you expect someone to pay up? One reason physicians are unhappy these days is that the definition of malpractice has changed. Malpractice is no longer defined as truly negligent or improper behavior. Now, a poor outcome alone triggers claims of "malpractice." The quality of the care may be irrelevant.

I have never been sued but I expect to be. The entire new generation of surgeons expects to be sued. Our elders tell us it's coming; it's just a matter of time. It doesn't matter how good you are or how carefully you practice. For that reason, I'm always trying to figure out which of my patients might be most likely to sue me. If it's really obvious (they gloat about the lawsuit they won against Dr. So-and-So when surgery wasn't everything they had dreamed it would be), then I'm likely to steer clear of them and recommend definitive treatment elsewhere. Most of the time, though, it's not so obvious, and you have to go with your gut. Unfair? Maybe. Paranoid? Not at all.

———

On the subject of risk, I have to include my two cents about medication. You might think that, because I am a surgeon and accustomed to "cutting people open" (as a few of my more eloquent patients have referred to it), the thought of simply prescribing a medication, or popping a pill myself, would give me no pause at all. Not so. I am a firm believer in the pithy statement: "Anything strong enough to help you is strong enough to hurt you." No treatment, at least no worthwhile treatment, comes without risk. Even natural supplements, if you take unnaturally large amounts, can have untoward effects, which is finally starting to dawn on the public. For that matter, placebo pills have been shown to have a whole host of vague "side effects" (headaches, dizziness, stomach upset), triggered either by the mere belief that the placebo is a real drug, or purely as a coincidence.

I've been lucky to be relatively healthy. I've never taken a pain medication stronger than Tylenol or Advil, never taken an antibiotic for anything, and have never been on any prescription medications. Aside from a garden variety multivitamin and extra calcium, I don't take anything. I'm wary of taking natural things in unnatural doses. I am amused by the burgeoning pseudo-medical chain stores that hawk supplements. My idea of maintaining a healthy diet is to make sure I sample from multiple different types of ethnic cuisine. The jury is still out on which culture has produced the best regimen for longevity, so I hedge my bets and include them all.

Surgery is visual and tangible. The risks are pretty concrete. Medications are a little more mysterious. There are plenty of medications that work wonders without us having a clear idea as to how or why they work. To me, that means that there are probably other things those drugs are doing that we may not expect. It would be unlikely for a drug to have one and only one effect on the body. That's not how the body works. One physiological mechanism can mediate numerous different functions. One natural chemical, blocked or enhanced by a certain drug, may have dozens of different targets. Those targets are probably not all figured out yet.

For that reason, I'm not surprised when scientists discover that a seemingly innocent medication, taken for years by thousands or millions of people, is found to be associated with a slightly higher risk of something scary and unexpected: blood clots, stroke, heart attack, excessive bleeding, sudden death, weight gain. This doesn't mean that I will shun all medication out of fear of adverse effects. I'll take a medication when I need it, when the time comes, if the benefits clearly outweigh the risks. But I won't expect to get something for nothing.

The thorniest medication questions arise with pregnant patients. What can you do for a pregnant woman with bad headaches or bad back pain? Is there any pain medication she can take or does she have to suffer? What if she has a seizure disorder? What antiepileptic medication is safe? What about if she gets an infection? Are antibiotics okay? The standard teaching pounded into my head was that *all* medications should be considered potentially risky, at least to the fetus. That seems a bit drastic, but not if you think about it for a second.

How would you go about doing a careful study to determine whether a medication is "safe"? You would have to recruit, say, one thousand pregnant women, early in their pregnancies, to participate in a medication trial. You'd want to pick relatively healthy nonsmoking, nondrinking women. Allow half of them to take a certain medication, and forbid the other half from touching it. Follow them through their entire pregnancies. See how their kids turn out. If 4 percent of the women on the medication have babies with birth defects, compared to only 2 percent in the other group, you'd have a nice study to publish in the medical literature and newspapers. ("Twice the risk of birth defects!" would be a great headline.) More meaningful warning labels could be put on medication bottles. This all sounds good—it sounds scientific—but who would volunteer to participate in such a study? No one. Now you can see why doctors don't have all the answers, and probably never will.

TEN

Emotion

I very rarely cry, but when I do it's all or nothing. I sob. And once the switch is flipped on, my eyes get red, my face gets blotchy, and it's obvious to anyone who sees me that I had been crying, even an hour later. My pathetic appearance then triggers the question, "Why were you crying?" just at the point where I have regained composure. This risks provoking the tears all over again. I am jealous of people who can cry just a little bit, gracefully, regaining their normal flesh tone right afterward. It's not such an ordeal for them.

I've cried on the job only once—but not in front of colleagues— and I've regretted it ever since. It wasn't because I was yelled at (not worth crying over) or because I was exhausted and overworked (not worth it either). It was because I had to deliver a death sentence, and I let my guard down.

A young healthy man in his late twenties, just a couple years younger than I was at the time, had a seizure while taking a shower one evening. His wife heard the loud thump and ran upstairs to find her husband unconscious, limbs flailing. I met him in the ER. By that time, he had regained full consciousness and was nearly back to his normal self. A CT scan of his head showed that a small patch of brain was slightly darker than normal. We knew this probably represented a stroke, a tumor, or an infection. But why would this young healthy person have a stroke or an infection? Those diagnoses were relegated to possibilities two and three—a very remote two and three. Brain tumors are more random. Risk factors are not required, and so we felt that this diagnosis was, unfortunately, number one on the list. Another more detailed scan the following morning convinced us that our hunch was probably right. He went to the OR for a biopsy and for removal of as much of the tumor as was safe to remove.

Based on his scans, we felt that the tumor was probably on the more benign end of the spectrum. It didn't really light up much with the intravenous contrast. The worst ones usually do. Going into the OR, we gave him the hope of a more benign tumor, and we had the same hope ourselves. We always leave an out, though, that "we can never be sure until we get a piece of it."

The pathologists usually require a couple days to come up with a final diagnosis. The day after surgery, the attending neurosurgeon on the case left town for a meeting. As I was the resident most involved with the patient's care, I was put in charge of looking after him and delivering the diagnosis. After a couple days had passed, I was on our usual evening rounds with the rest of the residents, and we stopped by the patient's room to check on him. He looked great. We all said hello and I promised to return by myself at the end of our rounds after I had spoken with the chief of pathology.

I finished my work and sat down at the nurses' station to call the pathologist. "It's not benign," he told me. "Definitely not benign. The entire department took a look at it and we all agreed. It's not what we were expecting." On the standard brain tumor grading scale of one through four, with four being the worst, this young man's tumor was a three. The problem is, a three eventually turns into a four, and once it's a four, life expectancy is usually measured in months rather than years. (Actually, the twos also tend to "devolve" into a more malignant grade, but it usually takes much longer.) I knew how this patient's life was going to change—and eventually end—and now I had to go tell him.

I walked down the hallway to his room. His door was open and a few of his good friends were visiting. His wife and new baby daughter were there, too. With all the joking and laughing, the mood was festive. The patient didn't look like a patient. Too cool to stay in his flimsy gown, he had taken it upon himself to change into his jeans, T-shirt, and baseball cap. The cap covered his scar. He shunned the hospital bed and was sitting in a chair by the window. He looked good enough to go home despite having just had brain surgery. He smiled at me and asked if I had a diagnosis for him. We sent his friends out to the hallway.

I sat down and delivered the news. I hinted at the ultimate implications of his diagnosis, but I didn't want to hit this too hard too soon. I wanted to give him some time to digest the shock of the unexpected. I looked at his wife, his infant daughter, and at him. He nodded his head, slowly, calmly. I wanted to provide them with some hope so I started, reflexively, to enumerate all the treatments he could receive that would give him the best possible chance. I reassured him that he was young and healthy, which would put him in a more favorable category.

I felt I had done enough talking at that point, so I stopped and sat in silence, a natural invitation for questions. I looked at the three of them. His wife was starting to cry, silently.

Then, without warning, I started to cry, too, then sob, interrupting the silence. My usual calm professional demeanor had broken down. I was struck by a harsh paradox: the vision of this young vibrant family sitting with me in the present, clashing with my knowledge of biology and how this tumor was about to change their lives. I could see the future too clearly.

The patient continued to look at me, stoically, nodding his head. He exhaled audibly and then thanked me. I didn't deserve much thanks, though. I worried that my unbridled outpouring of grief had wiped out any shred of hope. Chances are that if the surgeon is bawling, the prognosis is dismal. I calmed down, hugged his wife, and left the room, passing his friends in the hallway and looking downward to shield my face. I walked straight out to the hospital garage and drove right home.

Raymond Carver is one of my favorite authors. His minimalist style is vivid, disturbing, and memorable. After devouring all of his short stories, I was intrigued to come across a poem describing his thoughts upon receiving the fatal diagnosis of lung cancer. It gave me a sense of being on the other side of the conversation. It also made me wish he hadn't been such a heavy smoker. He died too young. The world would have enjoyed another collection of his stories.

What the Doctor Said
By Raymond Carver

He said it doesn't look good
he said it looks bad in fact real bad
he said I counted thirty-two of them on one lung before
I quit counting them
I said I'm glad I wouldn't want to know
about any more being there than that
he said are you a religious man do you kneel down
in forest groves and let yourself ask for help
when you come to a waterfall
mist blowing against your face and arms

do you stop and ask for understanding at those moments
I said not yet but I intend to start today
he said I'm real sorry he said
I wish I had some other kind of news to give you
I said Amen and he said something else
I didn't catch and not knowing what else to do
and not wanting him to have to repeat it
and me to have to fully digest it
I just looked at him
for a minute and he looked back it was then
I jumped up and shook hands with this man who'd just given me
something no one else on earth had ever given me
I may have even thanked him habit being so strong

The term "brain tumor" is a catchall term for just about any tumor that grows inside the head. In reality, only a subset of tumors arise from the brain itself. Most of the benign tumors—meningiomas—arise from the outer covering of the brain and tend to cause problems by indenting into the brain or pushing it aside. True tumors of the brain tissue itself—known as gliomas—are the most feared. These are also known as "primary" brain tumors. (Metastases, which originate from other cancers outside of the brain, such as lung cancer, are an entirely different entity.) Gliomas actually arise from the supporting glial cells of the brain rather than the neurons, or nerve cells. The most malignant form of glioma, a grade four, the glioblastoma multiforme (GBM), remains a thorn in the side of neurosurgeons. Despite sophisticated imagery, advanced operative techniques, chemotherapy, radiation, and millions of dollars in research, life expectancy is still abysmal, around one year on average.

The natural question most patients with a GBM ask is: Why did I get this? The truth as far as we know it, in the vast majority of cases, is that the tumor appeared for no particular reason. It is a random, unlucky, biological occurrence. A very small minority of patients are at higher risk because of a preexisting neurological syndrome, such as neurofibromatosis or tuberous sclerosis, or because of previous radiation to the head, but those risk factors account for only a small percentage. The rest are otherwise healthy, at least from a neurological perspective.

We'd love to be able to blame a certain chemical, work environment, home environment, bad habit, cell phone, or deity—we could work on prevention—but there's just not strong enough evidence against any of them. I'm confident that a GBM is not retribution for any sin or misspent life (it would probably affect more than just fifteen thousand people per year if it were). In short, a brain tumor is the fault of no person or thing. As with a deadly hurricane, nature is often both powerful and indifferent.

One thing that experts in our field do know after years of research is that there are a few stereotypical genetic mutations found within GBM tumor tissue. But again, no one knows what causes those genes to mutate in the first place. Once a particular cell has been tainted by mutant genetic material, it tends to multiply and proliferate like no other brain cells, creating a mass, invading once normal tissue, recruiting additional blood vessels required for rapid growth, and pushing aside anything in its way. When it attains a large enough size, the growing tumor may outstrip its own blood supply, leading to cell death (necrosis) in the center, which can then contribute to aggressive swelling.

The main problem with these tumors is that they don't have neat boundaries, regardless of how focused they may appear on scans. We can't remove the whole tumor because we can't see the whole thing. They're too diffuse. No matter how good a job you think you did, and no matter how great the scan looks after surgery, the tumor eventually comes back. It's frustrating as hell and a real blow to the surgical personality.

Consider what would happen if, when you try to open a bag of unpopped popcorn, the bag rips wide open, scattering the entire contents across the kitchen floor. A good-sized pile would end up right at your feet, and the kernels would radiate out from there. The relatively focused part of the mess would be easy to clean up. Over the next few days, though, you would be amazed to discover individual kernels that made it all the way under the dining room table or into the living room. That's what a glioblastoma is like. You can clean up the focused mess, but you know there are cells you can't see at first, far removed from the obvious focus.

Neurosurgeons have learned from their early predecessors. Decades ago, a few pioneering neurosurgeons, hoping to single-handedly cure patients of their glioblastoma, figured that all they had to do was remove a very large margin of normal-appearing brain tissue along with the tumor. That should prevent recurrence, right? They were so aggressive, in fact, that they would remove nearly the entire lobe of involved brain (the frontal lobe, for example), even if it meant causing a neurological deficit. That still didn't work. The tumor would come back in a different lobe, even on the other side of the brain. The glioblastoma could not be outsmarted, and it still can't, at least not yet. That hyperaggressive approach didn't last long.

Despite clear lessons from history, there is still plenty of controversy regarding the best surgical strategy for these tumors. Some surgeons will spend an extra hour or two in the OR trying to suck out every last bit of abnormal-appearing brain tissue, sending small samples from the margins to the pathologist as they go, until they get to "clean" margins.

Other surgeons argue this makes no sense and is a waste of time. Those surgeons aim only to "debulk" the tumor, with the simple goal of relieving any pressure and allowing some room in case the brain swells from the radiation. They acknowledge that a glioblastoma is not a "surgical disease" in that it can't be cured with an operation, and the extra effort to get as much out as possible doesn't confer a significant enough benefit (a few extra weeks, maybe?) over the less aggressive approach.

Some neurosurgeons dedicate their careers to these tumors, looking for alternative strategies to outwit them. Their work has led to some creative options, such as thin wafers of chemotherapy that can be left along the edges of the brain where the tumor was removed. Logic might dictate that this should greatly prolong the time to recurrence, but the results have been less than stellar. Clearly, a breakthrough solution will have to be radically different from the options we have now, and it's probably not going to involve surgery.

The most difficult decision is faced when the tumor comes back in full force after all treatments have already been exhausted. Do you make a second trip back to the OR? What about a third?

A close family member will often be the first to realize that a tumor has grown back. One woman explained to me how she knew it was time to bring her husband back into our clinic. We had operated on his brain tumor months earlier and he had been functioning relatively well. He was the host of a weekly polka hour on a local radio station. His wife always tuned in. One day in particular, the pauses between songs were too long and he even played the same song twice. She knew it was time, and she was right.

When neurosurgery residents get together in a pack, out of earshot of anyone else, the talk is often blunt. (A common saying such as "this guy is toast" seems unimaginably callous on the page, but it's said as an acknowledgment of the often harsh nature of reality, as in "life sucks" or "s—t happens." It's not meant to demean the individual in any way.) When the discussion turns philosophical, which it can on rare occasions, we wonder, for example, what *we* would do if handed the diagnosis of an imminently fatal brain tumor. Here is a sampling of the types of sentiments you might hear bounced around among the jaded:

"I'd be outta here, man. You wouldn't see me going through radiation, chemo, more surgery. I wouldn't prolong things; in and out of the hospital for my last few months; none of that crap. Give me a one-way ticket to Tahiti."

"I'd quit my job, go out with a bunch of beautiful women, sell my house, and blow through the rest of my cash. You gotta live it up while you can."

Then there's the more tempered, slightly less cynical, approach: "I don't know, I'd probably give radiation and chemo a try, but when it comes back, forget it. Look at Mr. So-and-so . . . three operations . . . that's nuts. We should have closed him with a zipper last time."

Although the words are irreverent, bold, and definitive, I wonder if the primitive, hardwired survival instinct would kick in and erase all thoughts of the classic "one-way ticket to Tahiti." Would we keep coming back for more, like so many of our patients, or would we remain true to our strategy of hedonistic nihilism? Would we really be different from anyone else, just because we've seen it all before, over and over again?

I met a patient with a glioblastoma who went against our usual expectations. He didn't use the word "fighter" or the phrase "do everything you can." The oncology fellow paged me at midnight on a Sunday. The patient had arrived at our hospital that afternoon from a small town a few hours away. His brother drove him. He was advised by his oncologist back home that he should consider undergoing chemotherapy at the university hospital. His glioblastoma was diagnosed three weeks earlier. It was multicentric—somewhat unusual—meaning that you could see more than one focus of tumor on his MRI.

At an outlying hospital, he had already had surgery to have the right-sided tumor debulked. The smaller tumor, on the left, was even more of a problem because it had a large cyst associated with it, near an area involved in speech. The patient had surgery on that side, as well, to have the cyst drained.

According to the patient's brother, he was originally reluctant to undergo surgery once the needle biopsy had confirmed the diagnosis. He eventually consented to a craniotomy, but decided that once was definitely enough. He knew that some people with a glioblastoma consent to have their head reopened once the tumor reappears, but that would not be for him. He understood that the tumor was incurable and preferred not to prolong the inevitable. His family was in agreement. The problem was, by the time I met the patient, he was not conscious enough to share this opinion. The brother spoke for him.

A scan at our hospital showed that the patient's cyst had reaccumulated a large amount of fluid from the surrounding tumor since the time it was drained, and must have reached a critical volume just before his admission that night, causing enough pressure to impair consciousness. During the car ride over, the patient's brother did note that he was a bit sleepy, but chalked it up to poor rest the night before. Within hours of his arrival, he had become sleepier to the point of obtundation: he only opened his eyes if you inflicted pain, he moaned but didn't speak, and he wouldn't follow any simple commands.

I sat down with the brother. Was the patient firm in not wanting any further surgery? Yes. What if we could improve his condition, at least temporarily, by redraining the cyst? Well, that would be surgery and he wouldn't want it. Did he write down any of his wishes in any formal way? No, unfortunately. The brother told me more about the patient. From the way he talked, I could tell that they weren't just brothers, but were good friends as well.

I learned that the patient was fifty, a biologist who specialized in fish, and was single. He loved the outdoors. He was a triathlete. It pained him to have his life slowly and progressively restricted by his tumor. It had started to affect his vision and strength. Because the tumor on the left was in an area critical for speech, he was experiencing worsening word-finding difficulties.

I explained to his brother that surgery to redrain the cyst would be straightforward, especially given the fact that there was a recent opening in the skull that we could work through. It would probably only take an hour or two, at most. He should wake up rapidly after the fluid was drained. I explained that his impaired consciousness was due to an increase in pressure from the fluid. If we could just drain the cyst, we could relieve the pressure and have him wake up. Otherwise, the end was very near.

The brother kept focusing on the longer term, though. He already understood the facts and wouldn't let me sugarcoat the situation. He had me acknowledge that the cyst would probably just reaccumulate again, despite repeat surgery, and that both tumors would continue to infiltrate the surrounding brain, even if slowed down by whatever new protocol was being tried at the time.

There was no convincing the brother to have the patient go for more surgery. He knew that the cyst was hastening death and he was okay with it. He knew his brother would accept this. At first, I was reluctant to let nature take its course, but my surgical tendencies softened after getting a clearer picture of this biologist's view of his own life. I called my attending and told him the story. He agreed to abide by the patient's and brother's wishes. We were dealing with an incurable tumor, after all. I went back to speak with the brother, left a long note in the chart, called the oncology fellow, and went to bed.

Early the next morning I received a page from the attending oncologist, yelling on the other end, irate as hell. "How could you just ignore this guy . . . the cyst? He's about to die! We were planning to start chemotherapy. You guys are going to take him to the OR. This is ridiculous!" Before calling to chastise me, the oncologist had phoned my attending and persuaded him to take the patient to the operating room. Neither had said more than a few words to the brother, who had slept all night by the patient's bedside. They gave him no choice.

I decided not to participate in his operation, a conscientious objector. Technically, all went well, the pressure was relieved, and the patient woke right up, with a plastic tube and bag hanging from his scalp to collect ongoing drainage from the cyst.

A few hours later, after the patient was sent from the recovery room to his hospital room, I was paged about him again. This time it was my own attending, the neurosurgeon who performed the operation. He requested that I stop by the patient's room to remove the drain and send him home. It turns out that after the effects of anesthesia wore off and the patient realized what had happened to him, he wanted out. He was upset that he had undergone another operation and was shocked to feel the plastic tube coming out of his head. He wanted the drain removed and wanted to go home. He had even decided against chemotherapy.

I found him in his room sitting up in a chair. His brother was standing and looking out the window. The patient was already dressed in his jeans and flannel shirt and was ready to go. His duffel bag was packed, by the door. The surgical catheter and drainage bag looked out of place—otherworldly—hanging from the scalp of this rugged outdoorsman. I was surprised he hadn't tried to remove it himself. I introduced myself and told him that I already felt like I knew him. He, of course, hadn't remembered meeting me at all.

In my brief interaction with this patient in his awake state, I sensed a strong resolve and a healthy realism rather than depression. I removed the drain and cinched down a single stitch around the exit site to prevent any fluid from leaking from his scalp. He pointed to the IV in his arm and asked whether or not I would mind taking it out, too. This is usually the nurse's job, but what the heck. He was in a hurry. The IV tubing had been secured to his hairy forearm by a large square sheet of clear sticky tape. "Just rip it off," he told me. "Better not to prolong the ordeal, if you know what I mean."

I once gave a talk to a group of undergraduate students who were learning about the brain. My role was to give a neurosurgical perspective. I wanted to have them do some reading beforehand, so I chose an interesting chapter out of a book written by a neurosurgeon about a patient of his with a glioblastoma (*Judith's Pavilion* by Marc

Flitter). I thought it was well written and provocative. I was hoping to stir up debate about the brain and neurosurgical interventions.

After introducing myself to the class I asked for their comments on the reading. The hand of a young cynic shot up. His comment went something like: "I don't know why anyone would want to have a career like that. I wouldn't choose it in a million years. Too depressing!" There was no argument, no debate, from the rest of the class. I hadn't anticipated quite so strong a visceral reaction. It made me realize that I had become almost too accustomed to death and devastation. It's a part of the job, and you can't cry over every patient. But is there a price to pay in getting "used to it"?

I have to admit that I sometimes wish I had become something like a handbag designer or a cookbook author (maybe not those exact things, but something along those lines). Those careers wouldn't involve the handing out of fatal diagnoses. I could enjoy a job with a little less gravity to it. My day-to-day would be more carefree, more creative, less serious.

I've even had other surgeons from different specialties divulge to me that they had, as medical students, strongly considered going into neurosurgery because of the brain and the neat technical aspects. "Too depressing" is almost always the reason they give for having decided against it.

So how can I save myself here? How can I justify tolerating this seemingly depressing job? (I am not a pure altruist and I don't think anyone is. The desire to "help humanity" can certainly go far, and that is still a primary reason, but there's got to be something in it for the helper.) Believe it or not, I have experienced at least one personal upside to seeing so much go so wrong. It sounds a bit hackneyed, but I have to admit that I have developed a distinct appreciation for everyday life. For many people, this requires some sort of personal near-death experience. You hear about these revelations all the time, about how someone didn't appreciate their life until they almost lost it.

I've been fortunate enough to borrow from everyone else's experience. I've seen people in every state of neurological decline and I've seen death, over and over again. And this makes me feel lucky about life, every day.

As I think about this, I have to admit that my appreciation for the everyday has become a well-entrenched part of me now. I probably don't need the constant reinforcement.

ELEVEN

Disturbing Deviations

I remember when I discovered that the lungs are actually beautiful, as internal organs can be, if you look at them with the right eye. I had seen real lungs before, the lungs of my cadaver in medical school, but those lungs were not the least bit attractive. They were inert, preserved, and cold. They were the lungs of a dead person who—I presumed—had probably been a smoker. I wasn't impressed.

Then, later in medical school, I scrubbed in on the open-heart case of a young girl with a heart defect. She was an otherwise perfectly normal girl except for this small defect and, luckily, a straightforward operation would be able to fix it. Although this was not my first time as witness to the human heart in action—I caught a glimpse of a routine bypass procedure when my dad took me on a tour of the ORs in his hospital once—it was the first time I was able to get a good look at living, breathing, human lungs. You might assume that the heart would upstage the lungs, but I'd say they're equally impressive.

The girl's lungs were pristine, pink, and glistening, and I could see them inflate in sync with the mechanical breaths of the ventilator. I was amazed. I could appreciate these lungs just as easily and as naturally as anyone can appreciate the skin of a baby. It's difficult to accurately describe the organ's look and delicate, airy, crepe-like feel. There is no other tissue or material in nature even remotely similar.

Midway through the case I had a sickening thought: What if this girl becomes a smoker? I imagined her, years later, as a college student with a cigarette, deliberately inhaling all the ingredients necessary for coaxing DNA into cancerous mutations, inhaling them directly into those thousands of fresh alveolar air sacs. I would be as devastated as her parents, maybe even more. If only the chest cavity were transparent, the beauty of the lungs would be exposed to everyone, and young women might be compelled to treat them as well as they treat their skin.

I think people enjoy their careers more if they find something that they work with to be visually appealing, like mathematicians who work with fractals or marine biologists who study jellyfish. Part of what I like about neurosurgery may seem superficial: I like the way the brain looks. The brain is, by far, the most complex and interesting organ of the human body (take note, guys, if you thought otherwise) and a feast for those who appreciate architectural detail. (It's not a gray blob by any means, if you take a good look. Actually, it's not even gray.) Unlike my revelation about the lungs, though, I can't remember when I first began to appreciate the brain. The brain is appealing not only in real life, but also as a line drawing in textbooks and as an image on scans—so much complexity packaged so neatly in a relatively small space.

Knowing what the normal brain should look like, it's also quite easy to have a visceral reaction to one that doesn't look quite right.

When I see a brain that deviates significantly from the norm, it can be jarring and even disturbing, despite the fact that I've developed a healthy degree of clinical detachment that I can deploy as needed. This is especially true when the brain is the brain of a child. In a fair world, all children would be born with beautiful, architecturally complete brains. But, given the randomness and indifference of nature, sometimes a brain just doesn't develop all the way, and we're left in the uncomfortable position of having to decide what to do.

The most abnormal brain I've ever seen belonged to an infant whose external appearance was frighteningly normal. His cute baby look was what made for such a sticky situation: the parents, relatives, and nurses saw a cuddly little boy who sucked on a bottle, cooed, kicked his legs, and even smiled. A basic and somewhat crude underlying fact, though, is that any infant's sucking, cooing, kicking, and smiling functions (or reflexes) require only the most rudimentary, nonthinking, parts of the human brain, the parts that are roughly similar to those found in much less sophisticated creatures, like reptiles. The overwhelming problem with this infant was that those primitive reptilian parts were the only parts that he had. The more highly evolved regions of the brain—the large cerebral hemispheres—the parts that allow us to develop conscious thought, experience emotion, and communicate with each other (in short, the parts that make us human), were completely absent. A cold, clinical, and involuntary question flashed through my mind on seeing this baby's scan: Is he really "human"? I kept the question to myself.

This condition, hydrancephaly, is rare. (And not to be confused with the more common hydrocephalus, or "water on the brain," which is easily treated and often compatible with a normal or nearly normal life.) Hearty survival is not expected in hydrancephaly, but it's hard to predict duration of survival, especially as we've become so expert at anticipating and intervening in the earliest hints of bodily breakdown, staving off what would otherwise be an early death. An infant who would promptly expire in less wealthy societies can be exquisitely maintained in ours by advanced technology, an abundant supply of skilled enthusiastic doctors, nurses, and therapists, and exacting parents. Such high-tech medical care is—normally—considered to be one of the key advantages of a wealthy and civilized society.

When I walked into the infant's hospital room for the first time, the lights were off and the shades down, in the middle of the day. The pediatric neurology team was crowded around the crib. The senior neurologist was holding a flashlight up against the baby's head, and all the residents and medical students stared at the resulting spectacle: a round, pinkish, glowing orb. Because most of the skull was filled with fluid and not brain, and a baby's scalp and skull are normally relatively thin, the light was able to pass right through, lighting up the head in an eerie display. This "transillumination" diagnostic technique is a very old and simple one, completely antiquated and unnecessary in this age of high-quality CT and MRI scans. Nonetheless, most doctors can't resist the temptation to try something that they've only read about in textbooks, and most doctors will never have the opportunity to examine an infant with hydrancephaly (especially given the ubiquity of prenatal ultrasound), so I probably would have placed the flashlight against the infant's head, too, if I hadn't happened to have caught the spectacle secondhand.

Here's why I was summoned to see this baby. His head was starting to grow faster than it should, out of proportion to the growth of his body. The normal cerebrospinal fluid created by the brain (the fluid filling up almost the entire intracranial cavity in this case) was not being absorbed properly. This was leading to an excessive accumulation of fluid. Because an infant's skull is somewhat malleable, such an unchecked accumulation leads to enlargement of the head. If left untreated, the head can become enormous, the child's skinny neck unable to lift it or move it, eventually leading to a downward spiral of pressure sores, full-thickness scalp breakdown, infection, sepsis, and death.

Scalp breakdown, of course, could be staved off for a while by keeping the child in the hospital, having the nurses adhere to a strict regimen of changing the head position at frequent and regular intervals, and by using creative padding techniques, similar to the fastidious care required to keep quadriplegic patients healthy and alive. Or, such care could be undertaken at home by hiring around-the-clock nurses. However, if the family were of more modest means, the task would probably be left to a nonworking parent, predictably the mother, who would then be at risk for mental breakdown within her confining and demanding world.

Another option, seemingly more humane, would be to implant a special shunting device into the baby's head so that fluid could be diverted directly from inside the head, through a tube tunneled underneath the skin of the neck and chest, all the way to the belly (a ventriculoperitoneal shunt). This would halt the excessive head growth and prevent scalp breakdown.

What to do? On the one hand, how can you allow this baby's head to continue growing larger and larger? On the other hand, how can you justify performing an invasive operation on a child with no future and only rudimentary brain structures? Is the goal really to prolong survival, to stretch out the emotional turmoil, for as long as possible? The child will certainly never be able to thank you for it, as he has no prospects for conscious awareness. The parents might thank you, but would their gratitude be misguided? And, because we have to think this way, too, is this a sensible way to allocate health care dollars?

During a neurosurgical residency, when a situation is so disturbing as to defy normal conversation, black humor is a natural, although admittedly juvenile, substitute. It's not directed at the patient or family (we're not cruel, really) but more at the awkward situation. The darkest of our black humor, while it would horrify any eavesdropper, is sure to lighten the situation and allow us to keep working, especially when we're overly exhausted and ready to snap. So, for example: What kind of want ad could this poor kid respond to when he grows up? "Looking for an individual who can reliably suck, cry, and kick his legs. Ability to smile also desirable." This type of banter, tossed around privately on rounds, is just another—albeit pathetic— way of saying, "Why the heck are we doing this?"

Strange, but true to the frenetic existence of a resident, I never found out what happened to this cute-on-the-outside little baby. I was involved in the initial consult but not the final decision. My four-month block on pediatric neurosurgery ended soon afterward and I was back to taking care of adults with architecturally complete brains who, among other things, decided against wearing a seat belt. In retrospect, I should have inquired as to what happened, but there were too many other pressing concerns, too many other strange cases to fill the day and night, and a constant stream of ever-renewing black humor to sweep me along, to keep me going.

During my time on the pediatric neurosurgery service of a busy academic center, I saw a wide spectrum of abnormal brain development. I also became facile with the complex names ascribed to each unfortunate anomaly. Many of the names have an unpleasant sound to them. I always felt awful for the parents who not only had to come to terms with the reality of their child's brain, but also had to learn, and repeat, the medical terms that went along with it: schizencephaly, pachygyria, holoprosencephaly, tuberous sclerosis (with "tubers" in the brain, for God's sake). There are so many names, a name for every conceivable anomaly or combination, many of which we may never see, that I had to create a large stack of flash cards in order to remember them all, at least temporarily, for my board exams. There is even a textbook of malformations that has been referred to—again, in our most juvenile but harmless way—as the "Little Shop of Horrors."

As residents, we would quiz each other with our homemade flash cards before our board exams. These were questions you'll never see on *Jeopardy!* (Resident #1, holding the card. "Port wine stain . . . seizures . . . railroad track calcifications . . . mental retardation . . . c'mon." Resident #2, with the answer: "Uhh . . . Sturge-Weber syndrome! Yeah!") For certain disorders, I didn't need a flash card because I had met at least one patient who exemplified all the features, and that was sufficient to fix it in my memory.

For example, a father brought his baby in, somewhat reluctantly, for consultation with one of the pediatric neurosurgeons. I sat in on the consultation. The infant had Apert syndrome, a congenital condition that includes multiple craniofacial abnormalities as well as syndactyly (fusion of the fingers), which are often surgically corrected, to the best of a surgical team's ability, when the baby is old enough and large enough to tolerate surgery and the resulting blood loss. In this condition, the sutures, or natural fissures, between different bones of the skull (such as between the frontal bones and parietal bones) fuse too early during growth of the head, leading to a misshapen cranium such as one that is abnormally broad or with an abnormally bulging forehead. Because of smaller than average orbits, the eyeballs appear to bulge. In addition, the "midface" from the bottom of the eyes to the upper jaw appears sunken in.

The father admitted that he and his wife had been shocked by their child's appearance at first, but then, over the ensuing weeks and months, had gotten quite used to it. They loved him, of course, as they would love any baby of theirs. Now, in fact, they had become so accustomed to his features that they felt he looked perfectly acceptable. They hated the thought of putting him through surgery. "He looks okay to me!" the father told the neurosurgeon, and then asked him, "What do you think?"

The next minute that passed was about as silent and as uncomfortable as a minute could be. The neurosurgeon had been rendered temporarily speechless, trying to figure out how to, at once, impart both honesty and compassion in his answer to an unexpectedly thorny question. I felt for everyone in the room—the father, the child, and the surgeon, in that order. After much discussion, the father was eventually convinced of the logic behind surgery, not only for significant cosmetic and social reasons, but also to allow for more room in the skull, so as to accommodate further brain growth. And with that, the father gave in.

The biggest mouthful, holoprosencephaly, is also known as arhinencephaly, which does not sound any better. It's a developmental defect of the midline brain structures (along the central dividing line) also typically associated with midline facial abnormalities. It sets in within the first few weeks of the fetus's life when structures of the face and brain are normally going through a cleavage or separation process. As with many other conditions, there is a broad range of affliction, from very mild with subtle defects to quite severe and lethal. Worst is the Cyclops form, with only one midline eye, and no separation between the two halves of the brain.

I will never forget one patient with this disorder, a longer-term survivor as she was seven or eight years old when I met her. Although her anatomical defects were relatively mild, she was still so severely mentally challenged that she was unable to communicate at all, and suffered from symptoms stemming from underdevelopment of the hypothalamus, an important structure that controls many of the "vegetative" functions of the human body. So, for instance, her body couldn't regulate its own temperature very well. I might approach her bedside to find her drenched in sweat despite the fact that her chubby form remained nearly immobile. On top of that, her sleep-wake cycles were always off. And, to add insult to injury, her body was starting to go through precocious puberty, which was distressing to everyone except her, as she remained not only immobile, but also unaware.

This poor girl would be in and out of the hospital for various problems at random intervals. With each hospital admission, multiple teams of doctors (pediatrics, neurosurgery, neurology, endocrinology, infectious disease) would all do their part to address the concern of the moment (clogged feeding tube? pneumonia? pubic hair?) and get her back to her mother, an exotic dancer whom I never met because she was never around. She was the type of mother who dropped off and picked up—definitely not the sleep-in-the-recliner-by-the-bedside type of mom.

Despite this girl's complex developmental anomalies, she was an "easy" patient, even described as "so pleasant!" by the parents of the child in the bed next to her. When I went in one morning to examine her, these parents, who had spent the night in the room with their own child, stopped me to say how impressed they were by their sweet roommate: "So quiet . . . never a peep . . . such an angel." I smiled and had to agree. It was not my place to expose the tragedy of her congenital misfortune to strangers.

I had the privilege of working with stellar pediatric neurosurgeons whose reputations brought patients in from all over the neighboring states and beyond. We saw the most complex, most bizarre, and most tragic of cases. Because modern medicine has become so good at treating every symptom, I was continually amazed at the children who were able to survive so long with so little brain function, with mothers, fathers, and hired help attending to every need and hauling around what amounted to miniature ICUs in their minivans. I became comfortable with conflicting emotions, thinking both "How touching!" and "How strange!" at the same time.

I had to perform a spinal tap on a child who had been neurologically devastated at birth. While I had the needle in his back, his mother mentioned that it was a good thing that she had forgotten to bring his Passe-Muir valve (an insert for a tracheostomy that allows a patient to talk). She explained that, this way, he would remain quiet through the spinal tap. I expressed some surprise that he actually had a Passe-Muir valve. I knew that his brain was not capable of speech or even thought. "No, you're right, he's completely nonverbal," she confirmed, very matter-of-fact and even smiling, "but he does make noises." This was not the first time I made a mental note to avoid making assumptions.

I enjoyed speaking to our visiting neurosurgery fellow from Turkey, whose amazement rivaled mine. "This would never happen in Turkey" was his common refrain. I never quite knew exactly what he meant by that, but I suspect it reflected some combination of wonder, respect, and horror.

At one point, late into my senior resident year, I was rudely awakened to the fact that I had become perhaps *too* jaded in dealing with the tragedies of neurological devastation. I had become overly accustomed to our clinics, going from exam room to exam room, one featuring a mother waiting quietly while she casually suctioned her daughter's tracheostomy, another featuring a twelve-year-old boy whose legs were so rigid and contorted that it was nearly impossible for his mother to change his diaper.

I walked into yet another examining room after checking the brief info in the patient's chart. It was a brand-new consult from out of town: eighteen years old, cerebral palsy, spasticity. Okay, okay, I've seen this before, I just need to get a good history before my attending walks in. Efficiency is key. I looked at the patient for a second: very skinny, special wheelchair, arms contracted, head support in place, mouth hung open. It was clear I wasn't going to get the story from him, so I turned to the parents, my back toward the patient, and started to take down the history. The mother's account went back eighteen years, recounting her pregnancy in detail. She was helpful but a bit long-winded, so I jumped in after a few minutes with some pointed questions. As a rule, I make an effort to let people finish their stories before I butt in, but sometimes I have to break my own rule. There were a number of other patients waiting and I didn't really need to know all the specific dietary details that people can't wait to tell a doctor about.

As I sat, dutifully recording the list of medications, allergies, and operations, my mentor walked in. I cringed. I was hoping to have at least the history done so I could present him with a nice summary. He sat down on the examining table, the only seat left in the cramped room. After introducing himself, he surveyed the compact scene—the patient, the parents—and then focused his gaze back on the patient. After what seemed like several, almost uncomfortably quiet, seconds, he looked the patient in the eye and asked: "So, when did you graduate from high school?" The young man's face lit up like I had no idea it could.

My mentor had noticed something I had missed. The patient was wearing a large high-school ring, so large that it looked a little silly on his bony finger. His body, far more than his mind, had borne the brunt of his cerebral palsy. He was a proud, beaming high school graduate. His mother pointed out the specialized computer, attached to his wheelchair, that helps him communicate. For the remainder of the visit I sat in the corner, duncelike, humbled by the enormity of this ring now staring me in the face.

———

This teenage patient reminded me, as an analogy to the trite "you can't judge a book by its cover," that you can't judge a person's intelligence by his outward appearance. But can you judge a person's intelligence by an examination of his brain? I was asked this question recently by a friend and my gut answer, never having studied the question in depth, was no. (This area of inquiry is more a Ph.D. concern than a clinical, practicing M.D. concern.) Obviously, I knew that certain conclusions could be made in the extreme cases—major congenital anomalies or devastating brain damage—but what about in the normal population? Can you look at the brain of, say, someone with an IQ of 90 and see any differences compared to the brain of someone who scored a 130?

Curious about whether my gut response had been correct, I looked around at what's been written on the topic recently and found out that the answer may actually be yes. I say "may" because I've learned not to put definitive faith in any one particular study, and the quality of the work has been variable.

For more than a century, various researchers have looked at overall brain volume and intelligence, and some have found a correlation between larger brains and greater intelligence. My feeling here is that pursuing the question in this very simplistic way is just not that interesting. You might be able to find some sort of trend, but my hunch is that there would be so many exceptions to the rule that the rule wouldn't be worth much. Einstein's brain, for example, was apparently slightly smaller than average. Looking at brain volume as a whole, or head size, is far too crude a measurement. I have to admit, though, that I may have a bias here as my own head is not particularly large. Anyway, consider the fact that the brain of a sperm whale is five to six times larger than ours.

What about approaching the question in a more detailed way? What if a pathologist were to hold an isolated brain at autopsy, rotate it around in his hands, look at it from every angle, and examine all the convolutions under a bright light with a magnifying glass? That wouldn't be good enough either. He still wouldn't be able to conclude: "I've got a genius here!"

You would have to look even closer, and know where to look. Much of the recent work has been based on detailed MRI (magnetic resonance imaging) studies of living volunteers. One study from California, for example, describes numerous focal areas of the cortex, in all four lobes of the brain, in which the presence of more gray matter was correlated with higher IQ.[1] Another group, in London, found similar correlations, but in different areas.[2]

Leaving aside the usual discrepancies between different research studies, a more important point is that examining structure alone has its limitations. Intelligence is a living, fluid entity probably better suited to study by *function*-based imaging techniques (in which images of the brain are obtained "in action" to see which areas "light up" while thinking.) One group, from South Korea, used functional

MRI (fMRI) to examine brain activation during reasoning tasks. They scanned two groups of subjects: one from the "national academy for gifted adolescents" and another from various local high schools.[3] They concluded that in the "gifted" students, the role of the posterior parietal areas was enhanced, based on its functions as part of the network that pertains to attention, reasoning, and memory.

Clearly, in order for the fragmented community of brain specialists to come to any meaningful conclusions about the physical brain and intelligence, much more needs to happen: existing studies need to be repeated or otherwise validated, the major questions need to be approached from multiple angles, and there needs to be communication across disciplines (the hard part).

For the time being, one attractive framework for thinking about the brain and intelligence focuses on flexibility as a key concept. Based on a review of numerous fMRI studies, scientists at Carnegie Mellon University believe that individual differences in intelligence may be based on the brain's ability to adapt to a changing environment, as in variable cognitive demands.[4] For example, people of superior intelligence may demonstrate more efficient use of brain resources, often requiring *less* cortical activation for a given task, exhibiting greater connectivity between brain regions, and having greater ability to recruit additional regions—quickly—as necessary.

All this, of course, has to be placed in the context of how important intelligence is in the first place. Isolated superior intelligence, buzzing around within the hard boundaries of an individual's skull, does not impress me so much as seeing what someone can accomplish with whatever resources they happen to have. And I know that there are plenty of so-called smart people in the world who aren't that smart at life, and that's not so easy to detect in the scanner.

———

The brain is such a complex organ, each one having to develop from scratch, I'm amazed that so many brains actually do develop so perfectly normally. After conception, not only does the brain have to escape all of the possible chinks in the genetic code, it then has to receive plenty of oxygen during the entire birthing process. Once a child's brain makes it past those hurdles, you would like to think it should be home free from then on. I wish. Other hurdles await. One child I took care of had a most unexpected roadblock to deal with: his own well-meaning mother. The disturbing deviations I saw in his brain were even more of a jolt than the congenital anomalies we see: his were entirely preventable.

I first met Tyler as the paramedics lifted him off the ambulance stretcher and onto the narrow table for an emergency scan of his head. He had just arrived in the pediatric ER by helicopter. I had become accustomed to meeting new people this way: the human being as a series of brain images attached to a name. Tyler was not aware of our impersonal introduction. He had recently slipped into a coma, and I, as the resident on call that Sunday afternoon, had to figure out why and what to do.

Tyler was a seven-year-old boy whose mother was recently divorced and trying to make a living as a masseuse at the local shopping mall. When Tyler came down with an aggressive ear infection, she drew upon her strong New Age sensibilities and treated the infection with carefully selected herbal remedies. Similar treatments had worked wonders for her kids in the past when they came down with colds or diarrhea (mostly self-limited viral illnesses, I would love to explain to her, and bound to "respond" to time alone). Her children, in fact, had been remarkably healthy, despite her refusal to have them immunized as infants.

What began as a typical childhood ear infection then progressed. Foul-smelling pus began to drain from his ear. She had great trust in her holistic remedies, but a growing apprehension led her to break down and consult with a pediatrician who prescribed a standard antibiotic regimen and urged prompt and complete treatment. (As an aside, while it is true that many ear infections are viral, and therefore do not require antibiotics, not all ear infections are the same.) However, still wary of commercial big-pharma antibiotics and, perhaps, of patriarchal prescriptions, Tyler's mother decided against filling the prescription and, instead, intensified her herbal remedies. She kept Tyler home from school and dutifully wiped away the pus that formed rivulets running down his neck.

During his mother's long weekend shifts at the mall, Tyler was entrusted to the care of his teenage brother. On a Sunday morning, several days now into his infection, Tyler woke up later than usual. Confused and lethargic, he barely made it downstairs, clinging to the railing, his eyes half-closed. His brother, roused from the couch in front of the television, found Tyler pale and clammy, babbling incoherently. He phoned his mother and, on her instruction, called the pediatrician, who met them at the suburban office. In the office, Tyler was found to be barely responsive and promptly had a seizure. He was rushed by ambulance to the nearest emergency room where he was intubated to sustain his breathing, and then transported by helicopter to our children's hospital.

I fixated on the brain images appearing one at a time on the console of the CT scanner. All four ventricles of Tyler's brain—the cavities filled with cerebrospinal fluid—were markedly enlarged, and his brain was clearly under considerable pressure. The story was clear: his unchecked bacterial ear infection had infiltrated through the inner ear, which lies within the temporal bone of the skull. From here, it is a short distance for bacteria to travel in order to reach the outer lining of the brain, the journey facilitated by the complete absence of antibiotics. Once this lining, known as the meninges, is violated, a raging bacterial meningitis ensues, clogging the normal circulation of cerebrospinal fluid and leading to extreme pressurization within the skull. This rarely happens in the developed world. Tyler was a cute little blond kid from the suburbs who reminded me a bit of my younger brother when he was that age. It was bizarre to think that his innocent brain was under assault from a third-world infectious complication.

Immediately after the CT scan, we wheeled him over to the intensive care unit, where I drilled a small hole through the top of his skull and inserted a thin drainage tube into one of the enlarged ventricles. Cloudy fluid spewed out of the end of the catheter with incredible force. I ordered potent broad-spectrum antibiotics to be administered stat through his IV. The herbal remedies were over. Meanwhile, Tyler's mother had just arrived at the hospital. The social worker recommended that I go out to the waiting room as soon as I could.

People often ask physicians how we handle the emotional stress of dealing with seriously ill children. What are our defense mechanisms? Do we practice a cool detachment? Do we shut off our emotions completely? Or do we go home at the end of the day and sob over a reheated dinner? I would say that cool detachment does come in handy on occasion, but the answer is really none of the above. The truth is, we are trained to do a job: recognize a problem, come up with a solution, and execute that plan. Our ability to actually *do* something protects us from what you might expect would be a chronic depressive state. We feed off the satisfaction of being able to help and we know that things would be worse if we didn't, or couldn't, do anything. For that reason, the experience of taking care of sick kids is much different from a hopeless walk through a pediatric ward as a visitor.

In Tyler's case, anger took hold before any sadness could even try to creep in. What kind of mother does this? I now had to go out and meet her, this all-knowing herbal shaman. What should I be thinking about her? What crime did she commit: child abuse or child neglect? How should I act toward her: professional, detached, and respectful, or should I promote the feelings of shame and guilt she deserves to feel?

I walked out of the ICU and opened the door of the small waiting room across the hall. She was there by herself. I shook her hand, introduced myself, and explained that I had just drilled a small hole in her son's head and inserted a tube into his brain. I had to relieve the intense pressure buildup from his aggressive meningitis, which was caused by an untreated ear infection. I added no allegations of abuse or neglect, no calls for guilt or remorse. I was, perhaps, more blunt than usual, but that's all I could muster. It turns out that, upon opening the door to the waiting room and taking a good look at her, sitting there by herself with the same look I had seen on so many other parents' faces, my anger was replaced by something else: pity.

Luckily, Tyler woke up within a few hours of the drain insertion. The antibiotics did their job and his meningitis cleared up within days. Still, at that point, he was not yet quite the same kid that he had been. He was a blunted version of the old Tyler, not as quick-witted or as cheerful, according to his family. As much as I am amazed by the normal brain of a child, though, I am equally amazed by its resilience once rendered abnormal by infection, trauma, or other catastrophe. Tyler was sent off to rehabilitation and I had high hopes for his future. Kids tend to bounce back over time. He may not become an A student, but I, for one, was satisfied knowing that he'd at least be a student.

An older, adult brain doesn't fare nearly as well under duress. His mother's brain, I was sure, had suffered a long-lasting injury of a different sort, and I couldn't predict how well she would bounce back.

TWELVE

Slices

Neurosurgeons have a love-hate relationship with the other professionals they depend on. Consider anesthesiologists. We can't do an
operation without them. On one hand, we enjoy the friendly banter
back and forth across the sterile drapes, and we like to tease them
about doing crossword puzzles at their posts while we're slaving away.
But on the other, such camaraderie doesn't prevent us from complaining bitterly when they delay our cases with the slow, tortured
placement of endotracheal tubes and central lines (even though a patient's challenging anatomy may be the main culprit). And, anesthesiologists serve as great scapegoats when a case doesn't go well, even
when they have done a fine job and have nothing to do with the complication. "Blame anesthesia," we might say to all present in the operating room when, for example, the total blood loss is announced at
the end of the case and is twice the expected amount.

In a similar fashion, failure to make a definitive diagnosis on a brain biopsy is often blamed on the neuropathologist, even though he has to make do with the quality of material supplied by the surgeon. Garbage in may mean garbage out. This concept is especially critical when it comes to the tiny brain specimens obtained by needle biopsy. There's not much for the pathologist to work with, so it's up to the surgeon to obtain the best possible sample.

As one of my mentors puts it, a brain biopsy is "two scared people separated by a needle." It involves the scared surgeon inserting a long, thin biopsy needle into the brain lesion of an even more scared patient through the smallest of stab wounds in the scalp and a hole in the skull no wider than a pencil eraser. It's one of the most minimally invasive forms of brain surgery. Very small lesions, deep within the brain—even within the brain stem—can be sampled with the aid of precise localizing techniques. Luckily, it's quite rare for something to go wrong—like bad bleeding—but we surgeons tend to worry about the rare.

During a brain biopsy, then, the mood in the OR can sometimes be a little tense. If the neuropathologist enters the room after his preliminary inspection of the specimen and requests more tissue, he may receive a sharklike glare from the surgeon no matter how politely the request is made.

After spending a couple years at the safer end of the needle, the time had come for me to play the much-abused soul receiving these tiny biopsy specimens of questionable quality in plastic petri dishes from the OR. During our required four-month rotation in neuropathology, we learn to be gentler to our neuropathology colleagues. We see that it's not always easy to make a diagnosis, especially when the surgeon doesn't give you much to work with. We learn the basis of staining techniques and microscopic diagnosis, and we participate in "brain cutting" autopsy sessions in which the specimens are much larger—the entire brain—and the surgeon is out of the picture.

The world of the neuropathologist is different from the world of the neurosurgeon. For better or for worse, there is no personal patient contact. This shields the neuropathologist from a certain type of stress pervasive in most other medical specialties. Also, because they have no direct patient contact, they can wear pretty much whatever they want. Some even get away with wearing jeans to work.

On a more serious note, the concept of hedging your bets is not necessarily frowned upon in neuropathology. A biopsy report might read something like: "could be A but cannot rule out B." This type of report drives a surgeon nuts. Neurosurgeons, on the other hand, are often ridiculed for being "frequently wrong but never in doubt." They don't do well with doubt or indecision.

Many neurosurgery residents merely tolerate this four-month requirement. They miss the OR, are bored by the pace, and feel a bit out of place in the nonsurgical milieu. I thoroughly enjoyed it. Often, I was intrigued by the aesthetics of what I saw under the microscope, the best of it truly worthy of display in a Chelsea gallery of abstract art. Equally, though, I loved the sane hours, which gave me plenty of time to immerse myself in popular culture, enjoy magazines and books again, see movies, check out the new restaurants I had missed, and spend time outdoors during daylight hours. As a cultural anthropology major in college I learned that leisure is the basis of culture, and I definitely felt more cultured during these leisurely four months, if only as part of popular culture and in catching up on who was dating whom. Neurosurgery residency has a way of making the outside world seem foreign at times, and this nice, easy rotation allowed me to feel like a native again.

Apart from enjoying the hours, hedonism, and aesthetics, I was lucky to work with a brilliant Cuban-born senior neuropathologist who maintained a fresher sense of wonder about the brain than anyone else I had worked with before, even though he was nearing retirement. In studying specimens with him under the multiheaded teaching microscope, I was delighted to hear his frequent "Look . . .

look how *be-zarrrre*!" as his eyebrows rose up above the rims of the eyepieces. He was a patient teacher. If I didn't understand what I was looking at, I could anticipate the "Look . . . look, I show you," as he explained a subtle finding for the third time. He would marvel at rosette formations, perfect mitotic figures, and double nuclei, and I would find myself marveling at all those little things with him.

There were many stories of his uncanny diagnostic skills, and it was hard to know which were apocryphal. Before starting the rotation, I was told by a senior resident that this guy could diagnose a meningioma (a benign brain tumor) "by smell alone." I found this claim hard to believe at first, but thought it plausible once I had gotten to know him. While I can't vouch for his olfactory skills, I can say he was an absolute master of the slick "touch prep" technique.[1] This method for rapid preliminary diagnosis was perfect for the tiny needle-based specimens as it involved simply touching the sample very lightly to a glass slide, thus preventing any artifact from freezing, crushing, or smearing the tissue as in other methods, and allowing the maximum amount of specimen to be left for permanent studies and special staining. He could glean all sorts of things from examining the individual cells that stuck to the glass. In contrast to some of his colleagues, he was more likely to deliver a solid, definitive diagnosis to the surgeon waiting in the OR with a needle in someone's brain: "This is A."

This professor had made his mark in the neuropathology world decades earlier for his work on a little parasite, an amoeba known as *Entamoeba histolytica*, a so-called free-living amoeba, which can invade the brain in certain cases. He had perfected his neatly packaged talk on the topic, notable mostly for the historic-appearing slides and the fact that this was a disease we would probably never see in our own country anymore with our sanitation standards and well-filtered water. Nonetheless, this tiny independent amoeba had helped secure his place in academia.

During my four months of learning and leisure, a two-year-old rhinoceros died at the local zoo. Nobody knew why. A full-body autopsy was performed at the zoo, but the brain was set aside and handed over to this senior sleuth for a specialized opinion. Right off the bat, we were puzzled by a clean slash through part of the cortex, and the zoo official explained in a matter-of-fact way that a chain saw and crowbar were required to open the skull, which was reportedly half a foot thick, so the process wasn't exactly delicate. With that, we had no further questions and got to work.

We took the fresh brain and weighed it: 650 grams, considerably less weighty than a human brain, which is usually around 1300 to 1400 grams, or about three pounds, in an adult. (To be fair, though, I don't know if a two-year-old rhino's brain would be considered adult size yet.) Neither of us had ever seen a rhino brain. We could roughly make out most of the structures—they were similar enough in location, but not exactly in size and shape—to a human brain. We were stuck, though, on the strange flaplike appendages protruding from part of the cerebellum, or hind part of the brain. There were no correlates in the human brain and I was inspired to find out what they were.

No one really kept tabs on my whereabouts during long tracks of the day, so I felt comfortable spending a few hours in the library looking for anything I could find on rhino brain anatomy, for anything that might link form to function. I found a few sparse references, but nothing detailed enough to answer my pressing question. In the end, the rhino case, although a fun diversion, proved ultimately frustrating: no clear answer as to cause of death, and no clues as to the what and why of those strange appendages.

In my animal brain research, though, which had become more wandering by the hour, I was enlightened to a tangential yet interesting curiosity regarding the brains of household pets: veterinarians will actually operate on brain tumors in cats and dogs, and the own-

ers of those cats and dogs will gladly shell out cash for these delicate
and expensive operations. (Are these the same humans who grumble
about their own co-pays and hospital bills?) I wondered if I should
have gone to vet school instead.

Near the end of my four-month rotation, our Cuban neuropathologist threw me a curveball. A group of residents was sitting around in
the microscope room, shooting the breeze about the brain, primates,
and evolution. I asked him his thoughts. He answered: "Evolution I
don't believe in. How could it be true? I don't believe." His answer
was as definitive as: "This is a glioblastoma. Period." I filed his creationist beliefs away as a personal quirk, and respected his willingness to speak his mind despite going against the grain. When it
comes to the fundamental questions, I guess you can't make assumptions, even when dealing with a scientist and fellow brain-lover.

Incidentally, some scientists believe that the human brain is still
evolving.[2] This theory is based on population studies of two genes,
both known to be somehow involved in brain size. A variant (or "al-
lele") of one of these genes, known as ASPM, likely emerged in the
human brain for the first time only about 5,800 years ago—relatively
recently on an evolutionary scale—and is already present in about
half of certain populations. This means that the new gene has spread
quite rapidly.

One controversy among many, though, is whether or not its
spread is truly a reflection of the natural selection of evolution. It
could possibly be via a more random process known as genetic drift,
whereby certain individuals, or populations, leave behind more offspring—thus, genes—by chance. Also, given the fact that human
culture, rather than brute biology, plays such a strong role in who
survives for how long, how well, and passes genes on to how many
offspring, the idea may be a stretch. Furthermore, we don't even
know how important this gene is. Although its mutated form can af-

fect brain size, the normal form may not. Apparently, the average
human brain size has not increased any further over the past two
hundred thousand years.

———

A couple years after my four-month rotation, I went back to my neuropathology friend's office to ask him for a favor. I needed to borrow
a few brain slices. I was preparing for a lecture about the brain that I
was going to give to an undergraduate class and I figured it would be
helpful to pass around the real thing as a hands-on demonstration of
look and feel. He was happy to oblige.

With his characteristic enthusiasm, he dropped everything,
sprang from his chair, and led me to a room down the hall where a
cardboard box contained numerous half-inch-thick slices of human
brains that had been prepared during autopsy. The slices had been
preserved in a formalin solution and were packaged in neat vacuum-
sealed plastic pouches, similar to what you might find in the meat
section of a grocery store. He had saved illustrative examples of a variety of different diseases: Alzheimer's, stroke, tumors, hemorrhage,
infections. It was a real treasure trove, for those inclined to view perfect examples of diseases as treasures.

He held up different slices one at a time and gauged my level of
interest in each, setting aside a short stack of the ones I liked best. I
only wanted four or five, as more than that would be a tight fit in my
book bag, and I knew I would have other things to carry across campus. I had reached my limit when he came across a final one that he
urged me to consider.

This last one looked familiar for some reason. Within this particular slice was a section of a brain tumor, an unusual brain tumor, and
I recalled having seen the exact tumor—size, shape, location—on a
patient's MRI scan not long before that. As a neurosurgery resident,
you develop a good visual memory for brain scans, and can remember them almost as you would a face. Often, when I see a patient in

follow-up in the office, an image of their scan will pop up in my mind as I greet them. (What I might say: "Hello, Mr. Garcia. How are you feeling?" What I might see: the real Mr. Garcia, and a flash-back of his scan from two weeks ago showing a small collection of blood in the right thalamus.)

My neuropathologist friend detected a puzzled look of recogni-tion on my face. "You know who this is . . . right?" He paused, wait-ing for me to complete the picture, to match the face with the brain. "Oh . . . yes," I answered, somewhat disturbed, and added the slice to my stack after some hesitation. I had to agree, it was a choice speci-men for teaching purposes, and I vowed to preserve its anonymity.

The next day, I left the hospital mid-morning and started walk-ing across campus on the way to give my lecture, the brain slices neatly arranged in my book bag. I walked past building after build-ing where university professors were teaching, holding court, en-lightening their students, and living their lives. I wondered how many were taking it all for granted.

I continued my walk past one building in particular, where one professor in particular, whose life had been cut way too short, had taught before his neurological symptoms crept up on him. I knew that it was *his* brain I had recognized, that I was now carrying in my book bag. His brain would continue to enlighten students more than his mind could have guessed.

I felt privileged to use his final, invaluable gift in the name of teaching, but I couldn't help but perseverate on a strange feeling similar to what I experience whenever I see Shel Silverstein's book *The Giving Tree*. The book is ostensibly a children's book, but proba-bly too weighty and disturbing for kids, judging from my own expe-rience of reading it for the first time. I don't even have to flip through the pages to get this feeling. Just catching a glimpse of the cover in a bookstore will do it.

In this book, a man uses his generous tree-friend for every imag-inable purpose, starting in his early years: picking apples and climbing the branches as child, lying in the shade and carving his initials (and girlfriend's) in the bark as a teen, cutting down the branches to help build a house as an adult, taking down the entire trunk to make a canoe and get away from it all in middle age, and, finally, coming back to rest on the stump as a frail, emotionally spent, elderly man. The tree keeps giving of itself, selflessly. It is the humble and humanlike personification of this tree, combined with the primitive line drawings, that practically has me in tears by the final page as the tree offers the only thing it has left—the stump of its former self—to the man who has taken everything else.

In working with the brain as an object, especially one dissected free of the body, I can't help but turn my thoughts to the philosophical once the anatomical and pathological requirements of the job are out of the way. Frank Lloyd Wright, when asked about his core religious beliefs, once answered something to the effect that he believed in "nature with a capital N." I like that answer. I know what he was getting at. I fundamentally believe in Nature, too, with the human brain as a key part of it.

To my way of thinking, there's really no role in this simple "religion" for the supernatural or the mystical. A pure nature-based outlook on life, I think, can be refreshingly straightforward: our brains, which make us who we are as individuals, function while we are alive. Once dead, brain function ceases, and the individual is gone. Figuratively, of course, the person does "live on" in a sense, but only because he is represented in the memories of the living and in whatever else he leaves behind: writing, photos, videos, artwork, donations, the DNA within his offspring, and so on. There is nothing mystical about this; no ethereal "soul" that floats around in the stratosphere or beyond.

I do have to admit that the supernatural elements of traditional religion offer a protective layer of comfort that a Nature proponent might miss out on. Take, for example, the idea of an "afterlife." Wouldn't it be comforting to think that you could mess things up in this life but make up for it all after death? Or, that you could make the tragic mistake of forgoing simple pleasures as a human on Earth, yet look forward to enjoying yourself more after burial or cremation?

Most people believe in religious teachings simply because they were brought up with them from an early age, not because they critically examined the fundamentals and concluded that they made sense. Culture and tradition often trump good common sense. From the viewpoint of a Nature-based believer, then, traditional religion can lead to a false hope or false comfort starting at an early age. Think of the ramifications. How many people sell themselves short on life because they expect great things after death? Life is not a dress rehearsal. You have to enjoy it, make the most of it, while your neurons are still buzzing with live connections. It's amazing how holding a human brain can emphasize these points, at least for me.

I don't want to give the wrong impression. I have the utmost respect for people's religious beliefs. I am, in fact, fascinated by those beliefs (in circumstances when it's not impolite to ask). I understand why religions were created and why they persist. The benefits of affiliating yourself with a religion are without question: a ready-made framework for morals, a welcoming social network, comfort in times of duress, and a repeating schedule of events (weekly worship, yearly holidays) that strengthen belief and bring order to one's life. Those are very nice benefits. The not-so-hidden downsides, though, can sometimes put a damper on these benefits: dividing humans along religious lines, encouraging war, discouraging marriage between otherwise perfectly compatible individuals, inhibiting free thought, and invoking guilt in those who stray from the flock.

I remember attending Sunday school as a youngster and feeling distinctly unsettled one day. I had just been taught, in no uncertain terms, that heaven was reserved for people who believe in Him (with a capital H). Most kids in the room felt pretty good about that. They were safe; they could laugh at the devil (lower-case D?) with impunity. I, on the other hand, felt awful. I asked the teacher about all the kids who happened to have been born in China. What about them? The teacher assured me that they, too, could be accepted into heaven as long as they found their way to Him as well.

Found their way? I imagined all the potential roadblocks to proper access and conversion: no churches in the province, no Bibles in the libraries, no soft-spoken Sunday school teachers to clue them in. I imagined a fiery subterranean hell with a sea of perplexed Chinese faces. I knew that just couldn't be right.

Ever since, I have remained the questioning sort, but I still thoroughly enjoy Christmas morning, Easter baskets, and religious wedding ceremonies. I'm not a spoilsport. I can appreciate the cultural traditions of any culture (particularly when good food is involved) and I do have a soft spot for the traditions I grew up with. The trappings of religion are not the issue as far as I'm concerned, as long as they're not taken too seriously. It's the dogma behind them that can sometimes be confining or divisive.

Sometimes, when I'm not exactly rational, like during the odd period between being awake and asleep, I worry that someone, somewhere, will force me to fill out the "religion" blank on a form, like on admission to a hospital, and won't accept N/A as an answer. "You have to pick one!" the persnickety lady behind the counter will say, handing the form back to me. I think about the world of possibilities. There are so many religions to choose from, how can there be just one "right" one? Could all the others be dead wrong?

Lutheran would be an easy choice for me because that is my family background, and nobody would question it. But which religion is really most in tune with my more brain-centric Nature beliefs? There's no particular option that's just right, but I think the basic ideas behind Buddhism are probably the most true to life and, one could argue, perhaps the most sophisticated. Along the way, as it evolved through different cultures, different forms of Buddhism picked up many colorful and supernatural elements that are a tad far-fetched (humans can't help it), like the idea of reincarnation, but the fundamentals are solid. (Unfortunately, I think the more colorful and supernatural elements are what can turn people off—so that they throw the baby out with the bathwater—and prevent them from taking a look at the basics, which are both ancient and very modern.)

A key idea in this philosophy (more a philosophy than a religion, per se, because no god is involved) is that suffering is a natural part of human life, and the only sure way to overcome suffering is to develop control over your own mind. That's what I like: it puts the locus of control in the individual, the individual mind, and not in any external power. That's a refreshing concept. You can enjoy the here and now because your thinking is clear, you don't have to look through the smoke of the mystical, and your mind is not the passive victim of whiplash in the turbulence of external events. The comfort you get from being in charge of your own happiness does not rely on any false hope.

I once toured the Asian gallery of an art museum and learned the meaning behind a common ancient Buddhist figure with dozens of small, uniform, individual clumps covering the head, almost giving the appearance of cornrows. The expert explained that these were meant to convey the numerous accessory brains of a highly enlightened individual. However far-fetched the idea of numerous accessory brains (clearly meant in a symbolic way), I was happy to see the brain represented so prominently in a religious icon. So many other religions downplay the brain, as if to deny that human minds, working collectively, are what created religion in the first place.

I volunteered to be a research subject several years ago, partly to help science, but equally to get a free MRI scan of my brain. As a neurosurgery resident and witness to a variety of intracranial catastrophes, I was curious to see if I might have any ticking time bombs of my own, or maybe a congenital cyst quietly taking up some room.

It wasn't part of the protocol to let the subjects take home their own pictures, but I begged the MRI tech, nicely.

My first thought, on seeing my own images, was relief to see everything in order, and nothing that didn't belong. My second thought was surprise and almost a twinge of disappointment at how it looked exactly like everyone else's normal brain. I'm not quite sure what I was expecting, but I must have felt subconsciously, however irrational and illogical, that the image of my brain might somehow be a reflection of me as an individual, like a face. In retrospect, the tech could have handed me pictures of anyone else's brain and I wouldn't have known the difference, except that I could identify my long, straight nose in the profile shot of my own.

Even logical minds are prone to illogical, mystical thinking sometimes. Luckily, though, it doesn't take much for me to reconfirm my more realistic, Nature-based outlook. Another day on call might present another brain at risk, with gradations of injury compromising gradations of the individual. Then instinct sets in: minimize the damage and maximize the individual. The brain is the mind is the individual, and it's our only hope. That's what I believe.

THIRTEEN

Fragments

Andy Goldsworthy appreciates the fragmentation of nature over time. I first saw Goldsworthy's work at the Museum of Contemporary Art San Diego, a small fine museum in La Jolla, California. I had been a fan of this artist's books for years. His books feature photos of his ephemeral creations, assembled with unprecedented creativity from objects in nature, like fall leaves carefully chosen along a color spectrum, fastened together in line with twigs or thorns, and allowed to float down a river. The leaf line eventually falls apart by the force of the river.

I was eager to see his work in person. The La Jolla exhibit featured organic objects, like large stones covered in a thick layer of clay, displayed right on the museum floor. The clay was applied while moist, and as these structures became drier and drier over days and weeks, they started to develop deep cracks along their surfaces. Over the period of the exhibit, the cracks extended all the way through, fragmenting the clay and further exposing the stones underneath. It was a haunting demonstration of impermanence.

Goldsworthy played to a similar theme in his various series of snowballs. He created larger-than-life snowballs with various other media, like sticks and stones, embedded within the snow, similar to bits of candy within a scoop of ice cream. He deposited these massive snowballs in various places, including in the middle of busy urban spaces. He photographed the melting process over time, recording not only the slow revealing and settling of the sticks and stones, which ended up in a neat pile on the ground, but also the puzzled looks of passersby on the street.

Much of Goldsworthy's art is not meant to last, except in photographs. This sets it apart from most other types of art I can think of, except for the Tibetan sand mandala, Christo's projects (like the Central Park Gates installation), and, I guess, urban graffiti, as these artists expect their work will get cleaned up, eventually. I first saw a sand mandala in college when a group of Tibetan monks was invited to create one at the Johnson Museum of Art at Cornell University. I've sought out several others since. Creating a sand mandala is a painstakingly detailed task, performed from memory by letting fine, colored sand sprinkle out of the end of a thin metal funnel onto a flat surface.

Museumgoers get to see not only the completed work, but even better, the work in progress. The sand sprinkling is carefully controlled by running a stick up and down along the rough surface of the funnel, held sideways, allowing the vibrations from the stick against the funnel to agitate the sand. These mandalas are quite large, and the monks have to work, logically, from the center out; they would end up messing up the edges with their robes if they worked from the outside in. They are very careful to preserve their fragile work during its creation, and most visitors naturally have the sense to keep a safe distance, in case they have to sneeze or cough, which might scatter the sand.

The final result is a masterpiece, an unbelievably intricate, multi-colored, circular design, full of symbols and complex geometry. Once the mandala is fully complete, the monks destroy it. They sweep up the sand and carry it out to the nearest natural body of water, in a ceremonial fashion. You have to enjoy the mandala while you can, just like everything else in life. Some things last longer than others, but nothing is permanent, and this is the whole point.

―――

Nature can fragment over time. The brain is part of nature. Degradation and disease are natural processes. Even so, it's hard to develop a calm acceptance of the fragmentation of the human mind. It's hard to find any beauty in it. The Japanese term for the beauty of things that are incomplete, worn out, or impermanent is *wabi-sabi*, and although I can appreciate the *wabi-sabi* nature of an old piece of furniture, with its chipped paint and random dents, or the natural, wrinkled face of a well-aged man or woman, I still have trouble with the wearing out or fragmenting of the brain, as most people do. I sometimes wonder if a good deal of suffering could be curtailed if a brain, like a Tibetan sand mandala, could be swept up and sent out to sea, in a thoughtful and respectful ceremonial fashion, before the intricate design has a chance to fragment too far, before it is degraded by the equivalent of drafts, clumsy feet, and curious hands.

As a neurosurgeon, I am often positioned at the end of the line. Similar to a police officer, I have often thought, who tends to meet people at their lowest ebb, as when a family relationship has degraded to the point where violence ensues, I often meet people at their lowest point, too, as when their minds have fragmented past the point of no return. I am then standing there, at the end of the line, asked to clean up what I can. It can be hard to shake the feeling of futility at times.

There are the fairly common scenarios, like the elderly woman with Alzheimer's disease who falls down the stairs and ends up in the emergency room with bleeding into the brain. How aggressive should I be for this poor woman whose mind has already been moth-eaten by disease? Should this be the final straw? What would she want me to do, if her mind were intact enough to offer a thoughtful opinion? It's at times like these that I might daydream of a career with a more upbeat focus.

(This reminds me of an idea I've had off and on during late-night treks to the ER. Everyone past a certain age—it's hard to be exact here and I'm not being ageist, really—should consider sleeping on a traditional Japanese-style futon, the real kind, frameless, right on the floor. I can't tell you how many ER visits for injuries to fragile parts—heads, necks, backs, limbs—could be prevented if simple falls out of bed were curtailed. It might also help maintain flexibility, with the daily getting up and down from the floor. One reason that some elderly can't get up after a fall is that they actually haven't been down on a floor for years!)

Difficult management decisions are made all the more taxing when family relations have fragmented along with the patient's brain. Understandably, it can be difficult to keep a relationship going strong when one person is barely there. I remember seeing a woman with advanced Alzheimer's disease who developed significant bleeding into an area of the brain that had already suffered a stroke. She was brought by ambulance to the ER, and the search was on for a responsible family member. The social worker eventually tracked down one of the daughters, who was kind enough to drive to the ER, check on her mother, and help us out.

I returned to the ER to meet the daughter. She was there with her somnolent mother, behind the curtain, and she had brought her own daughter as well. They had stopped at McDonald's on the way, and both were seated in plastic chairs at the bedside, dipping their fingers in and out of their paper French fry bags as I explained the gravity of the situation. I paused as they took the briefest of breaks to lick their fingers and sip on their Cokes. They continued, like metronomes, with their repetitive in-and-out pincer grasp motions as I explained that the bleeding could be fatal if nothing was done. They nodded their heads in understanding, their mouths full and lips glistening. I left the room to let them think things over and finish their fries. Typically, I would have shaken hands at this point, but I decided against it. I walked away wondering who, in this situation, would be grateful for the fruits of our labor: the patient? the family? no one?

In addition to Alzheimer's disease, serious mental illness is a particularly distressing sort of fragmentation. From my vantage point at the end of the line, the most extreme situations I've encountered have involved mentally ill patients. The situations have been extreme in many ways: the most gruesome presentations, the most marginalized members of society, the most convoluted ethics, and some of the longest hospital stays. After a patient has been in the hospital for more than a few weeks, we may refer to their stay as a "hostage crisis" on rounds, as in: "Mr. Doe, hostage crisis day number thirty-seven, self-inflicted gunshot wound to the head." It can be hard to get a person out of the hospital and placed in a facility when there's no insurance, no family, and no hope.

Gunshot wounds to the head—as accidents, attempted homicides, or suicide attempts among the desperately depressed—tend to fall into two opposite categories, like a dumbbell, with few cases falling into the gray zone of uncertainty: there are people who will survive and people who will not survive. It's usually easy to figure out which category a patient belongs to, and you can often tell the second they hit the ER. Breathing? Eyes open? Moaning? Moving? If none of the above then things look bleak. Their stat head CT seals the distinction. If a bullet has violated both sides of the brain, clipping off the deep vital structures (not just grazing across the tips of both frontal lobes, for example), then there's nothing that can be done, and the neurosurgeon's role is more counselor than surgeon.

Early on in my training, I learned a critical lesson in managing these patients: if a person is not going to make it, and you're sending him up to the ward for his final hours of existence, you still have to stitch up the bullet wounds in the scalp. Here's why: death is often precipitated by a buildup of pressure inside the skull. The pressure is due to the sheer volume of damaged, swollen brain tissue. Extreme pressure will eventually lead to one of two fatal events. Either blood flow to the brain will be shut off, or the brain stem—which harbors the control centers for breathing and heart rate—will become compressed and rendered nonfunctional. If the holes in the scalp remain wide open, the swollen brain "pulp" will naturally make its way out through these paths of least resistance, relieving the intracranial pressure, at least temporarily. A normalization of intracranial pressure can delay death.

One day on call, when I was still in the early stages of the learning curve with this clinical entity, I neglected to suture up the holes in a chronically depressed man who was nearly dead on arrival, but still breathing, his brain capable of coordinating normal rhythmic breathing but not much else. Out of ignorance, I elected to simply wrap his unsanitary head with gauze, as quickly and as neatly as possible, so the family could hold vigil at the bedside. I assumed he would expire before too late in the evening and I hoped that his family might be able to get some much-needed rest after a nightmare of a day. At that point, I was treating the family, not the patient, and thoughtfulness was the only thing I had to offer.

The following morning, on rounds, I was horrified to see that the patient's breathing was still going strong. A few family members had spent the night, half-asleep and distraught, on chairs and recliners in his room. I reported his condition to my attending, who asked me if I had sutured up the bullet holes. When I answered no, he put up his hands as if to say "What did you expect?" and recommended that I go back to the patient's room to finish my job.

I asked the family to leave for a few moments, for reasons I didn't feel obligated to explain, and sought out a nurse's help in gathering supplies, feeling like an idiot as I listed out the things I needed, all the things that I should have asked for the previous day.

I unwrapped the gauze and cringed on seeing that his hair had become matted down with brain tissue that had followed the simple principles of pressure dynamics. I stitched up both the entry and exit holes, the only truly thoughtful act I had performed for this family, and the patient expired a few hours later. Since then, I have vowed never to put a family through such prolonged torture. I've only made that mistake once.

Now take the opposite side of the dumbbell: an attempted suicide patient who *will* survive. This is an awkward and crushing situation. Survival is exactly what the patient did not want, but survival is what he gets, often a neurologically blunted survival, and often because of the position he held the gun in. I won't go into detail. I don't want this to be misconstrued as a how-to, but suffice it to say that the frontal lobes are quite forgiving, which I've said before.

––––––

The most disturbing mentally ill patient I ever took care of was sent to us "from the hills" of a neighboring state. During my training, I saw many unusual patients who came from those hills. Maybe it was because some of the communities out there were appallingly destitute. Illnesses sometimes presented themselves to us in quite advanced stages, having been neglected in their earlier stages due to ignorance or inadequate medical care. Add a fragmented mind to the extremes of socioeconomic misfortune, and the results can be disastrous.

I received the call around midnight. The story as told to me by the referring physician went like this. The patient was in his fifties. He had a long-standing psychiatric condition diagnosed as schizoaffective disorder (in the family of schizophrenia diagnoses, but with a significant mood component). A few years prior, a new small growth appeared on the right side of his forehead. He thought nothing of it. The growth enlarged, and he continued to ignore it.

Fast-forward a couple years, and the growth had taken up half of his forehead and had deepened its roots. When infection set into this large erosive lesion, the patient discovered at least a shred of decorum within himself and began to wrap a towel around his forehead, so as not to alarm anyone. Soon, the wrap encompassed not only his forehead but, as the tumor and infection spread, also his right eye. Still, he sought no medical attention. And, worse, no one had forced him to seek care. He did have a few distant relatives, but was otherwise an outcast.

Finally, after weeks of wrapping and rewrapping, he entered a convenience store and was stopped by the police for questioning. Other customers in the store had complained of a stench emanating from the towels around his head and face. The police took him aside, unwrapped the towels, and rushed the guy straight to the nearest emergency room.

A local surgeon at that hospital did a quick superficial debridement of the worst of the mess, including enucleation—removal of what remained of his worthless gelatinous right eye—and wrapped him back up until more definitive treatment could be undertaken. Infection and tumor had ravaged the entire right upper quadrant of his face and head. This was not a simple case for a small community hospital. The referring physician dictated a stat summary, gave me a call, and sent him over.

The patient arrived around two a.m. after a long ambulance ride. As he was wheeled down the hallway, I grabbed the envelope containing his medical records off of his stretcher. I read through the pages at the nurses' station while the paramedics and his nurse-to-be got him settled in. First, I reached for the brief operative note dictated by the surgeon. I fixated on statements that described the patient's wound, including—and I quote—"maggots of various stages of development." This was a first for me.

I looked through his blood test results and most of the values were flagged as abnormal. Clearly, whatever process was going on had affected not only his face, but also many of his bodily systems. He was not a healthy guy. I anticipated a hostage crisis.

I walked down the hallway to the patient's room. He was not talkative and neglected to say hello back to me, but was able to state his first name when I asked. The dirty towels I had heard about had been replaced by a wrap consisting of several loose layers of white surgical gauze. His flimsy hospital gown was draped over his body, and I could see that he was overly thin and bony, skeleton-like, clearly malnourished. I wondered if the tumor had spread throughout his body. He had no specific complaints to report and was vague and nonchalant regarding questions about his face and head. In his answers, he strung no more than three or four words together. He had no questions for me. There was no emotion. His remaining eye failed to make any eye contact.

I began the process of unwrapping his head so that I could assess what had to be done. I wore a double layer of gloves. His nurse, herself more quiet than usual, assisted from the other side of the bed. I was tired and felt a yawn come on but decided against opening my mouth. I was already repelled enough by the thought of filtering the ambient air through my nose.

The outer layers of the head wrap were clean and white, but the deeper layers were damp and faintly pus-stained. As I came upon the deepest layer, a small weak black fly escaped the wrap and landed on my arm. I shook it off and it landed on the bed, motionless. What we were left with, underneath the wrap, was a thick wad of gauze pads covering a large defect where his scalp used to be. I tried to lift up the corner of a pad near the center of his forehead. It was stuck. I could tell there was no bone underneath it. Through the unwrapping, the patient sat still and silent, staring ahead at the wall.

The nurse went out to the hall to grab some saline. We would have to soak these pads off. I didn't want to pull on them not knowing what they were stuck to. I poured saline over the entire matted wad, letting it dribble over his face. Once it was soaked, I was able to pull the wad off in one large piece. As I had suspected based on the notes, there was a good patch of skull that had been eroded through by a combination of tumor and infection. I examined the pads that came off. Small bits of necrotic gray and white matter were stuck to the deepest layer. With everything open to air, I recognized the unmistakable clarity of cerebrospinal fluid dripping down the side of his head.

The nurse and I looked up at each other and could find nothing appropriate to say. Then, looking back down at the patient, my eyes became fixated on a subtle movement. Was I hallucinating? I continued to watch until it became clear what I was seeing: a fat white maggot emerging from the man's frontal lobe. Feeling a wave of nausea, I retreated with the nurse into the hallway, leaving the patient alone with the parasite that had been feeding off of him for who knows how long.

I regained my composure. Knowing that I would have to awaken my attending, describe this patient over the phone, and come up with a plan, I forced myself to reenter the room. I made some quick measurements of the gaping defect and the extent of exposed brain. I glanced over the empty eye socket only briefly—that's not my field—but I knew that what I could see were the remnants of his extraocular muscles. As for the maggot, it was nowhere to be found. It may have headed back in. I rewrapped his head.

I needed a little more information before coming up with a plan. We wheeled him down to the CT scanner. I needed to know how the rest of his brain looked. How extensive would we have to get with this guy? How deep would our operation have to go? As the images appeared one-by-one on the monitor, I could see the radiology tech's eyes widen. We hadn't bothered to clue her in to the whole story. It was late. "I don't even want to ask" was her only comment.

I groaned when I saw the images. In addition to missing a portion of his face and head, he had an enormous brain abscess just below the exposed area of brain. He was in even worse shape than I had thought: a schizophrenic man missing part of his face, one eye, and with an abscess taking up most of his right frontal lobe. As one of my mentors likes to say at times likes these, "He's a winner!"

The following morning, I presented my new patient to the entire team of residents and showed them his scan. My plan was to take him to the operating room, remove the abscess and remaining right frontal lobe, and have our plastic surgeon buddies cover the whole defect in some creative way. My hopes of participating in this un-usual case were quickly dashed as my chief resident piped up: "Step aside. Looks like I'll be doing that case!" Even though I had done all the grunt work at two a.m., I had to respect our hierarchy. That's how things worked.

During the operation, I popped in and out of the room to check on the progress. The mood was festive. Black humor, of course, was bouncing around the room at regular intervals. My chief resident bellowed out to the scrub nurse, in a mock serious voice: "Give me the extra large suction tips, nurse, can't you see we've got maggots here!" The attending neurosurgeon wondered out loud, asking no one in particular: "So what did this guy say when he went to the bar-ber? 'Take a little off the top, but careful around the frontal lobe.' " And so on.

The plastic surgery team had a field day with this case, snapping photos as they went. They always enjoy a challenge and this was no simple nose job. They ended up transplanting a large flap of the pa-tient's own abdominal muscle and skin to cover the large defect, in-cluding patching right over the eye socket. Their work took hours, much longer than our crude brain-sucking exercise, and they were clearly quite proud of their masterpiece. I wondered how many slide shows this unfortunate patient would star in, at countless confer-ences, for years to come.

For the next three weeks, our team rounded on this man twice a day. We relegated the nitty-gritty of his care to the lowliest member of our team, the intern. Thanks to a lack of insight, he was placed in the very first room at the beginning of the hallway. Unsuspecting family members visiting patients farther down the hall were subject to a disturbing sight when the door to his room was left open. The view did not feature his good side, but instead showcased the handiwork of the plastic surgeons. There were several drainage tubes hanging out of the swollen fleshy construct that covered a good deal of his face. It wasn't pretty. He became our poster boy for the importance of early detection and treatment of skin lesions.

Despite our clearing the infection, forcing nourishment through his veins and into his stomach, and asking a psychiatrist to help us out, the patient never really perked up. We never had a real conversation with him. He never had any visitors. We checked out the rest of his body and there was no evidence that the tumor, which turned out to be a squamous cell cancer, had spread to other organs, despite its local aggression. He could survive in this state for quite a while. After numerous phone calls, reams of paperwork, initial rejections, and tens of thousands of dollars of free hospital care, our social worker found a nursing home willing to accept him, and we never saw him again.

In the end, what did we really accomplish with this patient? Does our work need to be appreciated by anyone in order to consider the effort worthwhile? I certainly detected no appreciation from the patient. He didn't even care that maggots had invaded his brain in the first place. (These maggots, by the way, however revolting, were to be thanked: they probably extended his life by eating away at the infection. He might have been in even worse shape if the natural world hadn't stepped in to help him out.) Given the baseline fragmentation of his mind and the additional destruction of brain function from infection, he seemed to have been rendered incapable of any appreciation, or perhaps any emotion, at all.

To make matters worse, whatever relatives he had certainly didn't care. They never visited or made contact. In cases like this, though, it's helpful not to think too much about questions of futility. Such philosophical musings risk inviting depression or inefficiency. Better just to fix the problem and move on to the next one. It's all part of being a service provider at the end of the line.

FOURTEEN

Controlled Trauma

As a child, I could easily spend an entire dreary Cleveland winter afternoon cleaning, organizing, and rearranging my bedroom, even though it was pretty neat to begin with. When perfection had been achieved, I would invite my parents in for a tour, encouraging them to open closets and drawers so they could admire the way I arranged my clothes or organized my stamp and coin collections. I took pride in attention to detail. I'm not sure if they were more proud or more worried. They certainly didn't promote such behavior in any way. There was no real cause for concern, though. My desire for neatness never reached the realm of the obsessive compulsive. I've seen obsessive compulsive—in the psych ward—and I'm quite confident that I'm simply neat.

I can even recall a brief period of time, as a very young kid, when I thought I might want to be a cleaning lady when I grew up. It was fun to clean things and then marvel over the results, so why not do what you love as a career? That thought quickly faded when it dawned on me that it wouldn't be as much fun to clean other people's rooms.

I have always valued simplicity, too, which I think goes hand in hand with neatness. At around the same age that I started cleaning my room as a hobby, I would pore over my mother's *Architectural Digest* magazines, looking for the stark white modern houses, the ones with wide open space and entire walls of glass. I also sought out the traditional Japanese houses with their clean, uncluttered aesthetic and neat tatami-mat rooms, tearing out the pages that I thought I could use as inspiration for my own future home. I would flip past anything too rococo, too ostentatious, or too golden.

I met my husband, Andrew, when I was a freshman and he was a sophomore in college. I remember him calling me up after one of our first "dates," if you can call them that, which consisted of our studying together in various empty classrooms around campus. We would spend almost as much time searching for the perfect classroom—walking all over the place at night, between the architecture school, the arts and sciences buildings, and the law school—as we would studying.

Over the phone, he asked me, in his characteristically straightforward way: "So, what's important to you?" I answered, in the abstract: "Simplicity and consistency." In retrospect, my rather Amish-sounding answer may have been a risky way to try to impress a guy. He took to it, though, and we continued to study together, which eventually led him to ask me out for dinner, a turning point in our theretofore ostensibly academic relationship. We went to the local college town Indian restaurant, Sangam, for curry and tandoori, and things have been both simple and consistent ever since. Years later, we still talk about the "curry effect" and its transformative powers in a relationship, the details of which I will leave up to the imagination. (The simplest relationship advice I can give, by the way, now over ten years into marriage: be honest, have fun.)

Lucky for me, Andrew was a natural neat freak, too, in all ways except for his personal style at the time, which was a carefully crafted messy look with the longer hair, untucked shirts, and few-day-old stubble—the exact look that college girls like myself went crazy over. His room in his cool off-campus apartment was a dead giveaway to his kindred neat-freak tendencies, though, with books on philosophy, biology, and poetry arranged just so on the shelves, and not a stray sock on the floor. His handwriting, even, was exquisite. I knew early on that this was a guy I could really like.

Some types of surgery are neater and cleaner than others. A slash-and-burn brain trauma case, for example, when minutes count, can be a mess: blood on the drapes, the shoes, everything; surgeons yelling, swearing; instruments flung aside after use. The scalp is flayed open in seconds, the bone flap drilled off and tossed over to the scrub nurse with what looks like reckless abandon. All motions are in hyperdrive until the blood clot is sucked out and the brain relaxed. At the end of the case, everything is closed up and the stress level settles down, but that's exactly when the complexity of care gets revved up: intracranial pressure needs to be monitored and tweaked, one or two of a myriad of possible infections might set in, other less lethal bodily injuries require attention, and the family is counseled through their shock and confusion, which takes time, care, repetition, and tact, day after day. The operation is the easy part.

Compare that to a type of neurosurgery that is so unlike surgery that its claim to inclusion within the world of operative procedures sometimes has to be defended. On my first day in the Gamma Knife suite as a junior resident, I fixated on a photo of our chairman standing next to the retro space-age-looking Gamma Knife unit, an early model with a spherical bulbous portion that the patient's head goes into, and a flat slide-like portion that their body lies on. He was wearing a handsome tuxedo and a subtle smirk. The caption read: "Civilized Neurosurgery." I knew I was in for a different experience from what I had been used to. I could clean the caked blood off my shoes and know that they would stay clean. Here was something that really appealed to my neat-freak tendencies. Had I found my niche? The two Gamma Knife surgeons in our department did have the neatest offices.

The Gamma Knife unit delivers a form of focused radiation to a specified target within the head. The procedure is called stereotactic radiosurgery. There is no cutting, no sucking, no blood. The radiation can be shaped in such a way as to treat an irregularly shaped tumor deep within the brain while leaving the surrounding tissues (brain, scalp, skull) relatively unaffected. The benefits over conventional radiation therapy are numerous: minimal to no radiation damage to the brain, no hair loss, and maximal dosage to the tumor. Even better, the radiation is delivered as a one-shot deal, not in multiple sessions over weeks. Add to all this that the patient has no incision, little pain, a simple overnight hospital stay, and no real recovery time and you have a pretty slick alternative to the old-fashioned "cut the head open" approach.

There are really only a few drawbacks. Accurate targeting of the lesion and shaping of the radiation requires that a four-pronged square metal frame be screwed to the patient's head during treatment (I mean, during surgery; the Gamma Knife surgeons are careful to avoid such a nonsurgical term as "treatment," especially when among other neurosurgeons). The application of this frame actually isn't as bad as it sounds, believe it or not, even though it looks like a torture device. It hurts going on but patients get used to it after a few minutes. ("It's like a tight hat," one of the surgeons likes to tell patients, which may or may not be a comforting analogy.)

Another problem is that the radiation often merely controls the tumor's growth rather than getting rid of it. (Tumor control may be all that's needed in many cases, but the idea does take some getting used to. Imagine a preoperative visit with the neurosurgeon: "Your tumor may never go away, but you'll be able to live with it just the way it is.") Also, stereotactic radiosurgery can only handle relatively small tumors, so only some patients are candidates in the first place.

Some traditional neurosurgeons scoffed at the Gamma Knife during my training. Partly, they may have felt a bit threatened by it, as patients steered clear of their offices and toward this bloodless option, and partly they remained skeptical of its efficacy. I was an easy convert, though, because I saw how pleasant the patient experience was, comparatively speaking. I would certainly choose it for myself in the right situation.

Some of the residents had fun taking little jabs at the Gamma Knife and stereotactic radiosurgery, precisely because of how tidy and nonsurgical the whole thing was. One of the more sardonically minded ones among us came up with a grading scale for the so-called pin site wounds, which were the small puncture sites (two in the forehead, two in the back of the head) left behind when the metal frame came off. These sites were simply covered with Band-Aids, which were removed the following morning before the patient went home. These pin holes healed just fine and were barely visible days later. On rare occasion, though, a pin site would ooze more than usual just as the frame was removed, requiring a single stitch. This was about as traumatic as the treatment (surgery) could get.

This Gamma Knife pin site grading scale, if I recall correctly, was a four-point scale as follows:

> GRADE 1: nontraumatic; Band-Aid only
> GRADE 2: steady finger pressure required
> GRADE 3: single stitch required
> GRADE 4: moribund

The "moribund" category was co-opted directly from other familiar, more hard-core neurosurgical grading scales in which moribund actually is a real and distinct possibility.

This pin site grading scale was recounted among the residents nu-merous times over a few-week time period, always eliciting uncon-trolled laughter at the word "moribund," until the whole thing fizzled out, landing in the rich wastebasket of other forgotten oral histories of neurosurgical residency.

Our hospital was known worldwide as a center of excellence for Gamma Knife work. We had the very first Gamma Knife in the country (the first in the world was developed by neurosurgeon Lars Leksell of Sweden in the 1960s) and we even had two of these multimillion-dollar units by the time of my residency. Along with holding courses for other U.S. neurosurgeons who were interested in participating in (or felt obligated to participate in) this bloodless trend, we had at least a couple foreign fellows learning the ropes at any given time so that they could spread the word across the world. Their camaraderie lightened the resident learning experience at times, as did the omnipresent flavored coffees and pleasant nurses, who knew they had some of the greatest nursing jobs around, with regular weekdays, normal hours, no mess, and easy patients. These foreign fellows were smart doctors but were usually new to the American experience and lingo, a deficit that tended to belie their true intelligence.

One of the Chinese fellows remained puzzled for months by a mysterious word used daily by the nurses in their instructions to pa-tients who were adjusting the position of their bodies within the Gamma Knife unit. After he got to know me he felt comfortable ask-ing, "What means this, 'scoot'?" And then a quick follow-up question, regarding a second mystery word commonly used in the same phrase as *scoot:* What means "butt"?

A Japanese fellow was equally puzzled when one of the nurses asked if he wanted to order a sandwich for lunch along with every-one else. He answered in the affirmative but then got stuck in com-ing up with an appropriate sandwich order, off the cuff and under pressure. The nurse prompted him: So what can I get for you, a turkey sandwich, ham sandwich, uh . . . knuckle sandwich? He asked for turkey but then turned to me for quiet clarification of the knuckle option that he had passed on.

This same fellow knew of Baskin-Robbins as "31," pronounced "tha-tee one," which is how it is referred to in Japan. The topic of Baskin-Robbins' ice cream came up during one of our cases (there's plenty of intermittent downtime during a Gamma Knife case when short, random, unimportant conversations can be held), and I informed him that we call it Baskin-Robbins, not 31, in the United States. In our country, I explained, the 31—a historical reference to the number of flavors—is more of an afterthought or a kind of subtitle. He paused for a moment and, hungry for more of such tutelage, asked what the true name for 7-Eleven was in the United States.

Despite the jokes, the soft ridicule, and the strangely nonsurgical feel of this surgery, the fact is that the Gamma Knife revolutionized neurosurgery, and remains a real boon for patients who otherwise might have had to endure a far less pleasant experience compared to lying in the Gamma Knife unit for an hour or two, listening to music of their choosing and exchanging pleasantries with the staff.

Given the tremendous advantages of treating a tumor with the Gamma Knife, the surgeons who performed stereotactic radiosurgery could be rather zealot-like in the promotion of their subspecialty within the world of neurosurgery, and justifiably so. They were amazingly prolific in the neurosurgery journals. Everything that had been written in the past about the efficacy of traditional surgery could be documented anew through the Gamma Knife lens, a different article for each different type of tumor or other category of disease. You could have one set of articles on technique and philosophy, another on early follow-up, and a later set on longer-term follow-up. The guys on the other side of the fence, though, could be equally zealous in their competitive promotion of maximal invasion. The grandstanding was most impressive at the national neurosurgery conventions.

An example that sticks out most in my mind was an opening talk that featured aggressive, complex, transfacial ("through the face") approaches to tumors at the base of the skull. It was delivered by a well-known neurosurgeon in the largest of the lecture halls. Every attendee received a pair of flimsy paper-framed 3-D glasses upon entering the room. This brought me back to the last time I wore 3-D glasses, which was for the 3-D version of the movie *Jaws* in the early 1980s. In this movie, the shark appears to swim into the audience with its mouth wide open, causing everyone to jerk their head backward like a fool, myself included. I wondered what might assault the audience during this show. Expectations ran high as the lights were dimmed, and this guy did not disappoint.

His lecture was a multimedia tour de force, most notable for the awesome 3-D skull models that demonstrated his technique, step-by-step, with pieces of the midface and underlying bone structure flying through the air toward the audience. These various human face parts were the hapless color-coded victims of every "trans" in the title of each surgical approach. The postop scans were undeniably impressive: not a single leftover speck of tumor in the deepest recesses of the skull—a classic neurosurgical "look what I can do" show-and-tell. We got to see the patients' faces, too, all put back together and smiling, after things had fully healed and they had returned to their human appearance. My skeptical side wondered about the patients he decided not to show us. When the lights came back on I looked around at the neurosurgeons seated around me. A few of the older private practice guys were shaking their heads as if to say, "You won't see me doing that kind of crazy stuff."

Neurosurgery is notable for the extreme variety of different "approaches" and the attention paid to them. When it comes to major surgery within the abdomen, for example, a nice long vertical incision right down the middle works fine for many of the internal organs. Once the abdomen is open, you can see and get to just about anything. A modified opening can be made depending on the target organ—a smaller one in the lower right-hand corner for the appendix—but it's still your basic linear incision, just of a different size, angle, and location along the topographically uninteresting surface of the abdomen.

Brain surgery is different. You can't gain access to everything through an invasive part, splitting the scalp and skull down the middle. The brain (and the blood-filled superior sagittal sinus) would be in the way most of the time. Many of our targets are hidden underneath the brain, not within it or on its surface. Furthermore, the head is round and irregular and there are critical surface structures—like the eyes and ears—that you have to avoid even if the target of interest is located underneath them. As a result, a panoply of creative and curvy incisions have been devised over the decades in order to sneak around or underneath.

The forehead is a key zone that we try not to violate, for cosmetic reasons, which forces us to make ridiculously long incisions (sometimes ear to ear) behind the hairline so that we can reflect and fold the forehead forward in order to reach, say, a small tumor just underneath it in the frontal lobe. In special cases, such as for big craniofacial reconstructions in kids, we may collaborate with plastic surgeons. They tend to be even more thoughtful than we are in fashioning a nice scalp incision. For example, rather than cutting across the scalp in a straight line behind the hairline, they may zigzag it all the way, giving it the appearance of having been cut with large pinking shears. When I first saw this, I wondered why they would go to the trouble. It takes longer and is more difficult to close at the end.

The attending plastic surgeon responded to my question as if the answer were obvious: What does the kid's hair look like when he gets out of a pool? If the incision simply goes straight across, the scar may be revealed with the linear parting of the wet hair. With the zigzag approach, the wet hair tends to cover the scar in a more natural pattern, not parting sharply along a dividing line, and the other kids at the pool may not notice it at all. Neurosurgeons don't think as much about poolside life, but I was touched by the concern, thinking back to my own awkward years as a kid, awkward enough without a large scalp incision and reconstructed cranium underneath.

The only real downside to collaborating with the scalp-friendly plastic surgeon is that we tend to worry a bit if we have to leave them alone in the room with the brain exposed. Our common refrain at times like these: be gentle; treat the brain as you would treat the skin. We all have our own nitpicky concerns.

Getting back to approaches, creativity is sometimes appreciated and sometimes not. The big buzzword these days is "minimally invasive," and some surgeons will go to great lengths to attract patients with claims that they are more minimally invasive than others. My husband and I were on a shuttle bus a few years ago, riding between an airport and the rental car place. We both independently took note of a businessman sitting across from us.

When we got off the bus, out of earshot of this gentleman, we turned to each other and said, quietly and simultaneously: "That guy got the 'batwing' incision." We felt terrible for him. The batwing incision is a rarely used approach to getting underneath the frontal lobes, just above the eyes. As an alternative to creating the aforementioned lengthy incision behind the hairline and freeing up the entire forehead all the way to the brow line (not very "minimally invasive"), alternative approaches are promoted from time to time by individual neurosurgeons hoping to push the frontiers.

The batwing incision, one such novel approach—again, almost never used, which added to our intrigue on the shuttle bus—involves creating a much smaller incision (more "minimally invasive") but one that is cosmetically challenging, within and between the eyebrows. Theoretically, most of the incision should be hidden once the eyebrows grow back, but it tends not to heal so seamlessly, leaving the patient with unusual looking eyebrows. Plus there's no way to hide the segment that crosses above the bridge of the nose. So, although it may be minimally invasive in one sense, it's maximally invasive in another. You have to be careful. A bigger incision can be the better choice, depending on the circumstances.

Long, complex, invasive skull base operations are the kinds of operations that always impress medical students. There's tons of interesting anatomy open to the air, so they get to see things they've only seen before in textbooks or a cadaver. It can be hard to watch at times, though, if you're not used to it. I once took a Ph.D. research colleague on a tour of our operating rooms. He had never observed any surgery and expressed a strong interest in catching a glimpse of a real brain, having studied brain function and brain images for years without ever seeing one. In one of the rooms, a skull base case was being performed, a combined effort between neurosurgery and otolaryngology (ear, nose, and throat, or ENT). We walked into the room just as one of the surgeons was gently tapping a mallet against a chisel into bone, just above and between the eyes, which were not covered by the drapes. The eyelids had been gently sutured shut to protect the eyes during surgery.

At the end of the tour I asked his impressions. He had been uncharacteristically quiet. He told me: "I could go outside right now and get hit by a bus, and I'd still be having a better day than that guy we just saw."

True, being in and around neurosurgery does have a way of inspiring a newfound appreciation for one's own health and luck. Still, his impression was amusing and I couldn't help laughing. All he saw was the height of the gore, the dramatic part, figuring that a pedestrian hit by a bus would be better off. He didn't get to see the guy a few days later, sitting up and asking when breakfast was served. You'd be amazed at how much can be done to the human body when necessary, with the human inside the body triumphing despite it all.

Surgery is trauma, but it's intelligent, controlled trauma, and it's done in the patient's best interest, despite appearances.

Similarly, I love when a family is thoroughly shocked and impressed upon seeing their loved one awake and talking immediately after brain surgery, as if they had expected the patient to emerge from surgery completely mute. Such low expectations have a way of fueling happier outcomes, even when things turn out just fine but no better than the surgeon had expected. These shocked responses, although amusing, are somehow endearing, and I certainly don't want to discourage them.

Going through my neurosurgery residency, I saw the gamut of options, from the most minimally invasive to the most maximally invasive. The neat and clean Gamma Knife approach certainly maintained its allure, but I decided that I wouldn't end up specializing in it. There wasn't any one particular niche within neurosurgery, in fact, that I could see myself focusing on exclusively, a reality that would become a small source of frustration later on. I just couldn't see myself super-specializing (as is customary in academia), narrowing things down so far in an already small world, confining myself to an even tighter box, even though that might have been the ideal career move.

Although I still sometimes dream of a pure white and glass house or even a perfectly traditional Japanese one (and who knows, I may still get one at some point), I am quite satisfied with the one we have now. It's over a hundred years old but newly restored on the inside. We've decorated it in an eclectic—but we think sophisticated—style, combining traditional colonial pieces from Connecticut with modern steel pieces from SoHo, and a thirty-dollar concrete Buddha head that appears to be ancient and expensive. We figured that if we couldn't commit to just one style, we would mix them together, carefully and selectively. The concept may sound messy but it works. And we always keep it neat, of course.

FIFTEEN

Traces of Thought

Four years into my neurosurgery residency, I started to get a little frustrated. I was dealing with the brain day in and day out, which is what I had asked for, true, but something was missing. I didn't have enough time or energy to really sink my teeth into the mind. A fascination with the mind is what got me interested in going down this road in the first place, but I was too busy sliding catheters through the cortex, drilling off bone flaps, and picking at tumors to think much about this ethereal by-product of the organ I had otherwise gotten to know so well.

Sure, in saving a brain we're also saving a mind. That's obvious. But the attention that we pay to the mind, per se, can be pretty superficial: this guy was in a coma and now he's awake and following commands; that guy is aphasic; this guy's memory is shot. Don't get me wrong; neurosurgeons are intelligent professionals who are good at what they do. It's mainly an issue of time management and focus. We just can't spend a lot of time mulling over the intricate nuances of the mind, dissecting apart all the different types of memory that a brain is capable of. Plus, it's not what we're the most expert at, except for a select few academics among us. We're not neuropsychologists.

The brain as a physical organ is what demands our attention most. Maybe I should have known better, but what other path could I have chosen? I knew I didn't want to be a psychiatrist or a neurologist or a basic scientist and the choices weren't infinite, at least not from the vantage point of a medical student traveling down a defined path. Neurosurgery seemed a better, although not perfect, career choice for me.

Luckily, as a part of our seven years of neurosurgery residency training, we are granted a full two years of research time. In my program, these were designated as the fifth and sixth years, just prior to the final year as chief resident. I was determined to remedy this lopsided physical/manual focus, and I became a sort of neurosurgeon-in-residence at the Center for Cognitive Brain Imaging at Carnegie Mellon University for one year. (I spent the second research year as a fellow in the small super-specialized field of epilepsy surgery, the subspecialty most directly concerned with cognition.)

The two codirectors of the center at the time, Dr. Marcel Just and Dr. Patricia Carpenter (a brilliant Ph.D. couple), focused their careers on studying higher-level cognition, including mental abilities that people value most highly, such as speech, comprehension, visual-spatial skills, memory, and decision making. They also study diseases of thought, diseases that neurosurgeons never deal with, like autism. That year seemed almost a guilty pleasure for me, not just because of the sensible hours and the fact that I never missed lunch, but more because I had the time to browse through journals—including non-neurosurgery journals—and to chat about ideas, and ideas are what get me fired up.

I was inspired to join their group after reading an article of theirs that was published in *Science*.[1] They were curious as to what it means for the brain to work harder. What's actually happening in the brain, for example, when you go from reading simple material to more complex material? In other words, how does the mind compensate for an increased workload? That was a recurring theme of their work. We pretty much understand how muscles respond to increasingly heavier weights or more repetitions, but how does the brain respond to a tougher workout?

They used a sophisticated brain imaging method, functional MRI (or fMRI), to look into this question. Functional MRI picks up on subtle changes in oxygen level in brain tissue. When a certain area of the brain is active, slightly more blood flows to that region, changing the oxygen level by just a couple percentage points. Those tiny differences in oxygen level reveal which parts of the brain are active or inactive at any given time. It's a reflection of the mind at work.

Normal research subjects or patients are asked to perform various mental tasks, like reading from a screen, while they are in the scanner. All sorts of clever tests can be devised to study all manners of thought or other types of brain activity, like coordinating movement. You've probably seen fMRI images before—in *Time* or *Newsweek* or *The New York Times*—with colorful splotches superimposed on anatomically detailed brain images. They've become quite popular in the press over the past few years, partly, I think, because they look so nice.

Their findings were intriguing. First, though, as background, the left hemisphere is well known to be dominant for speech (except in a small percentage of left-handers). Two of the major nodes in the speech network are Broca's area (in the left frontal lobe) and Wernicke's area (in the left temporal lobe). Most people don't think much about the right hemisphere when it comes to thinking about speech. This paper shows that we should.

What they found was that speech areas were recruited in a graded fashion depending upon the complexity of the task. In reading the simplest sentences, activation was seen in Broca's and Wernicke's areas. With increasingly complex sentences—dependent clauses, advanced vocabulary—a higher-volume activation of Broca's and Wernicke's was called upon. Then, for the most complex sentences, the mirror-image *right*-sided regions of Broca's and Wernicke's were recruited into action. In other words, the most complex reading called not only for a greater volume of brain use overall, but also for the recruitment of additional nodes in the network.

If you think about it, there are some interesting implications of this work. Most people who develop an aphasia (a speech or comprehension disturbance) after a stroke will eventually recover well enough to speak and read again. The brain does have an amazing ability to recover, at least partially. But what happens to their reserves? Are they capable of equally complex comprehension and equally complex speech compared to what they were capable of before the stroke? With a certain defined volume of their brain tissue rendered nonfunctional (i.e., dead), does recruitment falter with increased workload?

These sorts of insights inspire me to keep my mind as well tuned as possible, to use all the areas that can be used, so the reserves remain strong. I hate the thought of a particular region lying dormant for too long. Who knows how long you have before you turn the key and it doesn't start up again? I'm most at risk for this happening with my math skills, which were never that stellar to begin with, except for trigonometry in high school, which was more visual. I've become overly lazy with the calculator, to the point where it can take me longer than it should to figure out, in my mind, how much change I should be getting back at Starbucks when I get a tall iced chai. This is embarrassing. I clearly need more exercise.

I'll never forget my amazement at watching college-age research assistants at this center navigate through images of the brain on their computer screens. Their job was to help define the exact location of activation along the cortex, the convoluted surface of the brain. (By the way, the cortex is not always on the visible superficial surface of the brain. It folds in on itself in certain areas and also dives deep along the inner surface between the two hemispheres.) What I couldn't believe was how these students could point out and name any individual convolution (gyrus) or crevice (sulcus) so quickly— just as you might identify familiar streets on a road map of your hometown and think nothing of it.

My conclusion? These students were actually more facile with the detailed map of the cortical surface than most neurosurgeons. I, for one, had never before learned (or been taught) the names of every single sulcus, for example. I know the most important ones, of course—the most "eloquent" ones that we avoid violating at all costs—but not all the ones that we feel to be of lesser importance. They laughed when I told them this, and they clearly didn't believe me. I left it at that.

As far as I can tell, the lay public has conspired in telling the same joke when it comes to brain surgery. It goes something like this: *So, if you slip with the knife, is it like—there goes fourth grade?* Based on my rudimentary knowledge of how the brain/mind works, let me clarify a few things. First, we usually don't use any knives on the brain, except sometimes—if you really want to know—to prick the pia, which is the very thin, nearly invisible membrane adherent to the surface. Second, fourth grade (or any other particular grade for that matter) is not stored in any one spot. Memories aren't really "stored" in such a location-specific way, neatly and chronologically along a gyrus.

Memories are called up when needed via activation of the entire memory network across multiple regions of the brain, and if this process seems a bit mysterious and almost unbelievable, that's because it is. The more you know about how memory works and how much it is tied to the frontal lobes, emotion, and other such complexities (and how unlike a passive recording device the system is), the easier it is to see how memories can sometimes be faulty and how, for example, humans aren't always the most reliable witnesses in court or the most accurate in their autobiographies.

My husband and I can waste inordinate amounts of time in arguing over which one of us came up with a certain brilliant idea, exchanging accusations of faulty memory. On a trip to Europe I might say something like: "Aren't you glad I thought of visiting this little town?" Rather than a "yes" and maybe a "thank you" he'll counter with: "I like your revisionist history, but *I'm* actually the one who came up with the idea, remember?" Then I'll say, "But I'm the one who read about it." And he'll claim, "But the only reason you read about it in the first place was because I told you about it." And so on. Neither of us gives in. This makes me think that the memory network must also be tied strongly to whatever cortical regions control ego, but I haven't seen any studies on this.

Mysteries of the mind and brain abound, of course, and some may never be answered, but others are slowly unraveling with the help of functional imaging (pictures of the brain that reveal function, like functional MRI). Take the issue of blindness, and what goes on in the mind of a blind person. I've always been incredibly impressed by people who are completely blind, partly because of their Braille-reading skills and partly because some maintain the confidence and skill to maneuver around town on their own. I always wonder if I'd actually be able to do that, travel around on my own. Sometimes I conclude that the answer would be no.

One strike against me is that I have little inherent sense of direction. (Don't give me east-west-north-south directions unless I'm in a grid type of location that I'm familiar with, like the Upper East Side of Manhattan.) This deficit of mine is fertile grounds for teasing, again, by my husband, whose sense of direction rivals that of migratory birds. He marvels over how I can step out of a store in New York City, out onto the sidewalk, and start walking in the wrong direction. He likes to let me go for a while, walking alongside me, until he cracks up at my puzzled look as I pass shops that we had just gone into. I often rely more on recognizing landmarks than on a sense of direction to get me where I'm going. (Oh, there's that Japanese noodle place where we had dinner last month, now I know where we are.) As a blind person I wouldn't be able to rely on recognizing restaurants, and that's why I'd be in trouble.

I used to think my poor sense of direction was a real liability until I read that Harvey Cushing, the father of neurosurgery, shared the same deficit. It didn't stop him from operating on the brain and pioneering a whole new surgical specialty. I don't know, though, whether he worried about his potential as a blind person. Regardless, my husband probably would have had no problem laughing at his directional deficits either.

We used to live on the same street as a man who was blind. We never met him but we would see him walk down the sidewalk from time to time with his red-tipped white cane, no guide dog. One afternoon, during a heavy rainstorm, we decided to step outside our front door to watch the downpour. Just as we stepped out, we noticed that this man was making his way home, drenched, walking and tapping his cane back and forth at a faster clip than we had ever seen, turning sharp corners—onto our street and then onto his driveway—with rapid military-like precision. If he had moved any faster he would have been running. He must have had such confidence in the one-to-one correspondence between the well-traveled map in his mind and the external reality of the streets, he seemed to walk with a more self-assured stride than most sighted people.

Consider an interesting question: What happens to the visual cortex of a person who is blind? The visual cortex, a part of the brain in the occipital lobes (in the back of the head), receives visual input from the eyes. That's how it is stimulated. If there is no visual input, does that area remain completely fallow, nonfunctional, as logic would have you predict? Is it like a solar panel, of one purpose and useless without the sun?

This question has been studied, as I discovered in my browsing. A group of blind people who had been blind since early in life, with even no memory of vision, was studied.[2] They were asked to read Braille while in the scanner. For comparison, because reading Braille involves the tactile sense, they were also asked to perform other tactile tasks, some of which required fine discrimination (matching raised angles on a piece of paper) and others that required no specific discrimination skills (feeling a rough surface).

They found that the visual cortex in these blind individuals was strongly activated by Braille reading. It was also activated, but less so, by the other tactile tasks that required fine discrimination. It wasn't activated at all by the simple stimulus of feeling a rough surface.

A skeptic might interpret the data like this: Well, maybe the visual cortex is not as strictly specific for visual input as we thought; maybe it's involved in other sorts of fine discrimination, visual or not, and we just never looked for it. Maybe it's irrelevant that the subjects were blind. These scientists were smart, though, so they addressed the skeptic's question before he had a chance to ask it. They studied normal sighted subjects as well, using the same discrimination and nondiscrimination tasks (but not Braille reading, of course, because they didn't know how to read Braille). In these normal research subjects, the visual cortex did not "light up" at all, for any of these tasks. In fact, the visual cortex actually showed subtle decreased baseline activity as attention (and blood flow) was shunted to areas of the brain specifically in charge of touch sensations, an area known as the somatosensory cortex, in the front of the parietal lobe.

What this means is that because a blind person has no visual input, the parts of the brain normally involved in visual processing are freed up for other purposes, as when an old school shuts down and lies dormant for a while, but is then turned into something else, like apartments. The good real estate doesn't go to waste, especially in an active, dynamic community. (It might remain dormant in the setting of an unstimulated economy, though, to extend the metaphor.) The active learning and reading of Braille, then, allows new life to inhabit a dormant space in the brain.

A skeptic can be persistent, though, and he might look at a study like this and assume: okay, so the visual cortex is activated by reading Braille—granted, that is surprising—but the activation is probably just some sort of epiphenomenon, not really functionally relevant to the reading process.

Wrong again. Another study was designed to prove that activation of the visual cortex was critical in blind Braille readers, and not just some sort of an afterthought.[3] They used transcranial magnetic stimulation (TMS)—a cool and somewhat sneaky technique that can stimulate the brain via a magnetic force through the scalp and skull—to sort this out. While blind individuals were reading Braille out loud, they placed the TMS device over various parts of the head (with the subjects' permission, of course). When the TMS device was placed over the occipital lobes specifically, where the visual cortex resides, subjects made a significant number of errors in their reading and would even perceive missing, faded, or extra dots in the Braille text.

These studies are a great demonstration of what is called "cross-modal plasticity"—a type of plasticity or flexibility of the human brain in which one part of the cortex can shift its purpose, when necessary, to accept a different type of sensory input. It also may explain the common observation that people who are blind tend to have certain other heightened sensitivities. With one sensory modality shut down, there is more brain volume available for other sensory functions (hearing, touch, smell). There is similar evidence that the auditory cortex is used for sign language in the deaf.

(By the way, TMS has been used in other very interesting ways. An article in *The New York Times Magazine* described its experimental use in enhancing creativity. A nonartist was asked to draw a dog before, during, and after stimulation, and the differences between the drawings were quite remarkable, with the "stimulated" drawings clearly having more of a lifelike animated feel compared to the unskilled childlike baseline.[4])

One more thought on the topic of blindness. Apparently, Braille can be read using just one finger, the index finger, or three fingers at a time, the middle three. Unlike memory, the representation of tactile sense along different parts of the body actually *is* typically neatly defined along a gyrus, in the somatosensory cortex. There is a different area that represents the face, hands, feet, and so on, mapped out as a so-called homunculus: a map of the human body in which certain more sensitive parts—index finger, lips—are given greater representation, as opposed to the less finely sensitive back, for example.

What's amazing here is that, for blind people who use all three fingers simultaneously, the cortical map of these fingers becomes somewhat "smeared" and overlapping rather than distinct.[5] As a result, if you were to touch one of the three fingers individually, such a person might not be able to tell you which of the three is being touched. All three fingers have fused, in a sense, both in terms of their representation along the cortex and in terms of their sensory function, working as one unit.

In talking to patients, we sometimes use a term like "silent region" or "relatively silent region" to make them feel more comfortable with the fact that we'll have to go through their right frontal lobe, for example, to accomplish our surgical goals, as in placing a shunt catheter into the ventricle. (Some patients fixate on this concern excessively whereas others don't even question it—"whatever you have to do, Doc, no problem.") In truth, there *are* areas that are "relatively silent," but mainly in the sense that injury to those areas is well compensated functionally, even to the point of being undetectable in everyday living.

If you brought up the idea of a "silent" area to scientists like Drs. Just or Carpenter, which I have, they would argue. Functional brain imaging studies haven't found any areas that are nonfunctional, regions that never "light up." In other words, you can forget the old adage that we use only 10 percent of our brains. I remember hearing that "fact" from a misinformed friend in elementary school, but I don't know where the idea originated. Regardless, you can delete the 10 percent fallacy from your brain and use the room for something else.

The study of traditionally mushy topics, like emotion—strongly tied to the frontal lobes—used to be considered a bit risky for serious neuroscientists to take on, but this area of inquiry has definitely been gaining in credibility over the past ten years or so. (Studies on this topic may help explain some of the 90 percent of the brain that some of you had assumed to be useless.) I feel particularly lucky when my random browsing—even in the more popular press—lands me good material in this department.

Interestingly, because Buddhism is so strongly focused on cultivating the mind, the study of well-seasoned Buddhist monks has been quite fruitful. Put aside the robes, the chanting, and the shaved heads for a moment. Look past whatever cultural trappings might distract (or attract) you. Consider these guys Olympic athletes—gold medal athletes of the mind. They practice meditation techniques and other mind-related exercises just as athletes practice a sport. As a result, they can develop enviable strengths that put the average harried, anxiety-prone, modern brain to shame.

There is evidence from brain imaging studies, for example, that the ratio of activation between the right and left frontal lobes (specifically the prefrontal cortex, in front of the motor and premotor areas that control movement) is tied to happiness and a sense of well-being. Too much right-sided activation is associated with a tendency toward depression and anxiety. This ratio goes a long way in explaining the concept of a set point for emotion, different for different individuals, in which we may temporarily swing one way or the other depending on the circumstances (marriage versus divorce), but tend to settle back to our baselines before too long.

This knowledge, by itself, may be cause for depression, especially in those who fear their ratio may be shifted to the right, but the good news is that there is also evidence that this ratio can be improved with practice (no drugs!). Richard Davidson, a neuroscientist at the University of Wisconsin, had the chance to study not only nearly a couple hundred regular volunteers, but also a senior Tibetan monk.[6] The monk's brain activation was the most heavily shifted to the left out of all the other research subjects. With this knowledge as inspiration, non-monk volunteers were taught brain-strengthening meditation techniques similar to those used by Tibetan monks, and were able to shift their ratios over time, concurrently noting that they felt better overall, less prone to being derailed or unraveled by the common annoyances of daily life.

Most people I know, though, would complain that they have neither the time nor the energy required to train their brains to become stronger, happier, smarter. Spending several hours a week meditating or playing mind games in the name of self-improvement sounds like a good idea, but "it's not going to happen," most would say.

What we really need are the right implants . . .

After a year of poring through journals, pondering the complexities of cognition with smart Ph.D.s, and daydreaming, I believe that I can see the future of brain surgery. It's exciting and a bit frightening. But these thoughts will have to wait. I'm heading back to the OR.

SIXTEEN

Focus

The patient on our operating room table was just a baby. She was the sixteen-month-old daughter of Latin American immigrants. For months, she had been having anywhere from one to ten seizures a day. That's why we were looking at the surface of her brain, preparing to remove a small piece of it. We needed to get rid of the abnormal brain tissue that had become an irritant, the focus of her epilepsy.

The problem was, we couldn't tell exactly which small piece to remove. We had a clear idea going in—based on her EEG (brainwave test) and the abnormal spot we could see on her MRI—but things weren't as clear as we expected once we elevated the overlying portion of skull and dura and took a look at her cortex. It looked absolutely normal.

The other problem was that we had to be exact here, even more exact than usual. The abnormality was small and in a very unfortunate location: the part of the motor cortex that controls fine motor functions of the hand. This is not a "relatively silent" region. This is high-priced real estate. Luckily, given her extreme youth and the enviable plasticity of an infant brain, we had high hopes for a nice recovery. Still, we wanted to be as precise as possible in removing this thing, so as to leave all the surrounding brain perfectly intact for her future.

We bring out all our technology in a case like this, no cutting corners, so we had our high-tech 3-D image-guidance system up and running. This system, which is registered to correspond to a patient's preoperative MRI with near millimeter accuracy, is designed to give us extra confidence in homing in on the right spot. That was another problem, though. The system seemed to be off for some reason—several millimeters off—and our typical troubleshooting tricks couldn't correct the glitch. Its inaccuracy rendered it useless to us. We couldn't trust it. Technical difficulties like this do arise on occasion, as every neurosurgeon knows, and are generally more of an annoyance than a crisis. Usually, we can see the abnormality with our eyes. This wasn't a usual case.

So we continued to stare. We went back and forth between the convoluted gyral surface of her brain and the MRI images hanging up on the light box. The images teased us: they revealed the bright spot in her brain that we knew was the culprit, the seizure focus. If we could just match them up . . . her brain to the images . . . this thing was supposed to be right there on the surface. I knew that sometimes a lesion could be hiding just a millimeter or two underneath the surface, and you can't always discern that level of distinction from a scan. Maybe that was the issue.

We were confident we'd be able to figure it out, though. This wasn't a crisis, and it wasn't rocket science. But we were, maybe, just a little worried. The room was more still and quiet than an OR should be.

Then, a breakthrough. The senior professor of neurosurgery standing next to me, teaching me the ropes, a surgeon I look up to, not only because of his height but more because of his expertise, had an idea. He put his finger on the brain and gently slid it around over the surface. He looked at me, nodded, and smiled. (A surgical mask does not hide a smile.) Putting your fingers on or in the brain is typically frowned upon. It's sort of a low-class move akin to eating non-finger food with your fingers in a classy restaurant. This professor frowned upon it, too, in general, but he was savvy enough to know exactly when to go against etiquette.

I followed his lead, running my own finger over the pristine surface of this baby's brain. The lesion was obvious to the touch, a hardened knot embedded within soft normal brain. Then we laughed, partly because we were now safe to consider the whole situation funny, and partly because we were relieved. We got back to work and everything went smoothly from then on. And the baby did beautifully, as babies tend to do.

I spent my second research year—my sixth year of residency—as a fellow in epilepsy surgery. There are only a few places in the country that offer formal fellowship training in this subspecialty field. Despite the small number of spots, though, and the fact that the field is a fascinating one, the competition is not overly intense. A choice spine surgery fellowship tends to be the hot item now, partly because of all the new and improved "toys" available for implantation in the spine to treat painful degenerative conditions, and partly because these new implants—as a realist would be quick to point out—can enhance not only a patient's life, but also the bottom line. In the future, epilepsy surgery may become more popular as new technology brings new possibilities, implants, procedures, and demand, similar to the transformation that occurred in spine surgery.

I was lucky to work with one of the real masters and thought leaders in the field. He had a kind of dual identity: as a neurosurgeon and as a Harley-Davidson biker. He acted more like the former but looked a bit of both. Tall and thin, bald, and with a prominent white beard and mustache, he wore cowboy boots to work every day. In the winter he wore a black leather vest underneath his suit jacket and, in the warmer months, suspenders. He used an old-fashioned pocket watch rather than a wristwatch. He did wear ties, and had a soft spot for ones adorned with seahorses because the hippocampus, a part of the brain often involved in epilepsy, was named after the scientific term for this sea creature whose tail curls in on itself in a similar way.

This professor was the consummate academic neurosurgeon—like neurosurgery founder Harvey Cushing—intellectually curious and entrenched in basic research. He was busy and in demand, even overcommitted at times: seeing patients in the clinic, operating, running a department, serving on committees, working on the next research grant, attending international conferences. He didn't get to ride his motorcycle (or "donorcycle" as we tend to call them in the head injury business) as much as he wanted or as much as you might have assumed. Somehow, though, he kept it all together.

Epilepsy surgery intrigued me for many reasons. For one, you can sometimes actually cure people of their epilepsy. I still find that incredible. "Cure" is not a word we get to use often enough in medicine. Also, it is the branch of neurosurgery most concerned with the mind and cognition, which is what led me to neurosurgery in the first place. Collaborations with neuropsychologists are routine, for every case, to assess language, memory, and other functions. In removing a seizure focus, you want to preserve the patient's mind as much as possible—obviously—and a detailed array of cognitive tests can help us figure out if that's going to be possible. Maybe this small subspecialty, then, would be my ticket to lifelong career satisfaction, with its blend of interesting science, interesting surgery, and devotion to the mind. For one full year, it would be my sole concentration.

Epilepsy surgery is the purest form of brain surgery. By that I mean that you're actually operating on the brain itself, not around the brain, underneath it, or through it, as is otherwise often the case.

In epilepsy, there's something wrong with the substance of the brain, usually an area of the cortex, and in surgical candidates we have to try to fix it, by finding and removing the abnormal region. I also like the imaging. Epilepsy surgeons (and their very close allies, epilepsy neurologists) rely on multiple different types of images—MRI, PET (positron emission tomography), SPECT (single photon emission computed tomography)—and try to correlate the imaging findings to every other piece of data, hoping to localize the seizure focus. As a very visual person, I enjoy examining interesting images, especially when the brain is the subject.

Most intriguing and rewarding, though, were the patients. As a fellow concentrating only on epilepsy surgery candidates, I got to know them better than I had during other years of my training, which were more frenetic. Each patient went through a complicated, time-consuming, and often tedious workup. I often saw them over multiple visits, before and after surgery. It was lower volume but higher intensity—more my style.

Some patients were highly functional, with careers and families of their own. At the other extreme, some were on the fringes of society or even institutionalized, a combination of seizures and mental deficiency having devastated their lives and even their families' lives. It is a peculiarity of our profession that we do some of the most high-tech, interesting, labor-intensive, and expensive work on some of the most compromised individuals. Thinking back on our epilepsy conferences, I wonder in what other scenarios would you find a dozen intelligent people (neurosurgeons, neurologists, fellows, neuropsychologists, neuroradiologists, nurses) gathered around a conference table, focusing great time, energy, and resources on a single patient who may never comprehend what is being done to them? What might be accomplished, in addition, if the same group lent some of their collective brainpower to, say, improving public education or homeland security?

I remember many patients. There was the pleasant, heavyset

teenager who traveled all the way from Turkey with his family for epilepsy surgery. Just like the baby with the hidden lesion, this patient also happened to have an abnormality in the area of the brain that controls the hand and arm. His English was decent but he had a habit of referring to his arm as "my armie," which I found endearing.

Based on his MRI, we knew he probably had a condition called focal cortical dysplasia, which means that a portion of his cortex did not form properly during his brain development. Otherwise, he was a perfectly normal and healthy teenager. Cortical dysplasia is often associated with seizures. His seizures were mainly focal in nature, involving only his hand and arm on one side, rarely progressing to the type of seizure that would involve the entire body, the kind associated with unconsciousness. In other words, he remained fully alert during his seizures.

We did his surgery "awake," which means we performed an "awake craniotomy." This sounds a bit more dramatic than it really is because the patient is awake for only a certain critical portion of the operation. He is not awake during the scalp incision or the drilling of the bone flap or the closure at the end—the potentially painful parts. The anesthesiologist allows the anesthetic medications to wear off once the brain is already exposed, and puts the patient back to sleep after we complete the necessary awake testing. The brain itself has no pain fibers (a detail that I think is now common knowledge), so the procedure is better tolerated than you might expect.

With our young Turkish patient awake, brain exposed, we talked to him as we systematically explored the abnormal area of his cortex with a fine two-pronged electrical stimulator. We figured out which parts of his brain corresponded to his hand and arm and, as we feared and expected, the dysplastic cortex was mixed right in. This would be a balancing act: remove enough tissue to decrease his seizure frequency but not so much that his hand and arm function would be compromised. A complete cure of his epilepsy would be a real bonus; we weren't counting on it.

At the tail end of the stimulation the patient started to have one of his typical seizures, right there on the operating room table. We were alerted to this by the anesthesiologist, as we were on the other side of the drapes, unable to see the patient's hand. "Cold irrigation, please," my mentor asked of the scrub nurse, who promptly handed over a syringe full of cold, clear saline solution. Usually, we're careful to use nice warm saline on the brain and the rest of the body, but cold saline on the brain can actually stop a seizure in its tracks—a detail that is probably not common knowledge.

With the seizure aborted, we continued working. We removed the small area that we felt we could get away with removing, and did multiple subpial transections, or MSTs, on the portion that we had to leave in place. This is a technique in which we make small fine cuts in the surface of the brain. The theory here is that these cuts can prevent a seizure from spreading horizontally through the cortex, while allowing function within the remaining vertical columns of brain tissue. It sounds good, but it tends not to work out quite so perfectly. In this young man's case, his seizures did decrease in magnitude and frequency, but were not cured, and he was left with a slight hand weakness that required some time to recover from. All in all, though, the trip from Turkey and the neurological trade-off were definitely worth it.

When brain tissue is removed in a case like this, it's not treated like just some inflamed gallbladder or gangrenous toe, walked over to the pathology department in a plastic bag in the hands of a candy striper. This particular professor would send out a call to his entire research posse and, within minutes, fully fledged Ph.D.s toting special containers, complex preservative solutions, and dry ice would show up in the OR, wearing flimsy disposable white "bunny suits" over their street clothes. They would line up, quietly and respectfully, and would each receive a small piece for their own special projects. It always reminded me of Halloween. "Trick or treat!" was the only thing missing.

In the future, a patient such as our Turkish teenager might be fitted with some form of an electrical stimulator implanted over the cortex that would predict and then prevent a seizure, obviating the need for any actual brain removal, which, if I take a step back and think about it, does seem a bit primitive.

There is one type of electrical stimulator for epilepsy already on the market. It's called a vagus nerve stimulator. The thin, coiled, insulated leads are implanted around the vagus nerve in the neck, which lies sandwiched between the jugular vein and the carotid artery. The nerve originates in the brain stem, so when a segment of the nerve in the neck is stimulated, electrical impulses are able to travel in a retrograde fashion up to the brain, with the nerve as a conduit. A little battery pack is implanted under the collarbone area, just like a pacemaker. Typically, the device is not curative, but it can function as a valuable adjunct to medication, without the typical side effects of medication.

My mentor was skilled at talking to patients, including kids. He really wanted them to understand the plan. Once, we sat down in the clinic with a seven year old boy who was scheduled to undergo implantation of one of these vagus nerve stimulators the following week. His seizures were mainly "complex partial" in nature. This type of seizure—the most common type that we operate for—can be quite bizarre to watch because the patient loses awareness but not necessarily the ability to do things. A patient may, for example, remain staring with his eyes open but demonstrate strange "automatisms" like repetitively smacking his lips and blinking. When it ends, the patient typically awakens with no memory of what just occurred.

This boy's mother explained to us that if he had a seizure while eating, he would continue eating but would stuff everything in his mouth, reflexively, maniacally, including chicken bones or whatever else happened to be on the plate. We were hoping to curb that sort of thing.

We sat down with the young boy and his parents. The parents listened in from the sidelines, as their son was the one we were really trying to connect with. He was earnest, bright, and curious. He listened intently to the surgeon's entire explanation: an incision in the neck, another one under the collarbone, thin electrodes up here, a small battery down there, everything should go smoothly, a little pain medication for a few days, overnight for one night in the hospital. The boy nodded and took it all in.

Then he got to look at the actual device, one that we kept for teaching purposes. He fingered the thin electrodes and the battery pack. He turned it over in his hands. "Do you have any questions?" the professor asked. The boy continued to feel the electrodes and then paused.

"Yes," he said, looking up. "The 'lectrodes . . . they're a little bit sticky." (They were coated with a thin layer of insulation, and had been touched by countless other hands.) It was more of a statement than a question. He waited for the response.

"Yes. You're absolutely right. They are a little sticky," my mentor replied.

And with that simple confirmation, the boy was satisfied, ready to go forward with surgery, and the consent form was handed over to his parents to sign on his behalf. Given the same operation, different patients will fixate on different concerns. One thing I've learned is that regardless of what the surgeon assumes is most important, what a patient thinks is most important *is* what's most important.

The personalities of certain patients with chronic epilepsy make me wonder: How much is the person, and how much is the disease? In other words, how much of what I'm seeing in a person is a result of their seizures versus how much is just their inherent personality and abilities? This is often impossible to answer, which is why I like the question. There is one patient I can remember, though, in which the answer did become quite clear, but only after surgery.

Consider the life of a child growing up with medically refractory epilepsy (not well responsive to medication). For one, she is probably on more than one medication, maybe even three or more. Each medication has its own potential side effect profile, perhaps a vague and generalized slowing or fatigue.

She may have seizures during school. It may be a challenge to make and keep friends. Seizures interrupt the day, the lessons, and the learning. She falls behind. She has seizures at home, and this interrupts homework and family life. So, her grades are a problem, her social life is lacking, and her family is frustrated. She has poor self-esteem and a worrisome future. Certainly, the seizures and the medications can be blamed for much of this. What would her abilities and personality have been like if she had never developed epilepsy?

We evaluated such a girl in her early teens. She had had poorly controlled epilepsy for years, with her seizures occurring mainly at night, interrupting sleep (nocturnal seizures). She had a twin sister who was completely normal, who did not have epilepsy. Her twin was considered by everyone to be the older sister, and her family even referred to them as older and younger. Even though they were the same age, the patient was far less mature, more impulsive, and a much worse student compared to her sister. She had a short fuse and was a challenge to her parents and teachers.

Her original brain scans, performed years prior, revealed nothing abnormal. Medications were her only hope but none worked well enough. She tried numerous drug combinations. Her neurologist didn't want to give up on her, though, so he referred her to the academic A-team for further testing. She was put through an extensive evaluation: inpatient monitoring over several days, continuous EEG, updated brain imaging, and neuropsychological testing.

Imaging technology continues to improve. The MRI that we obtained during her evaluation was of significantly better quality than the one she had had years prior. We agreed that her old MRI *did* look normal, but something wasn't quite right on her new, high-resolution one. There was a hint of something amiss in the right frontal lobe, but the finding was extremely subtle, and maybe even an artifact, maybe not even a true finding. The team persisted and ordered a specialized variant of an MRI in which a coil is placed over one part of the head, accentuating only that one area while the rest of the brain fades into less detail. That was the ticket. There *was* something in the right frontal lobe, probably a very small focus of cortical dysplasia.

We took her to the OR, found that little spot, and took it out.

Fast-forward a few months to a follow-up visit in the office. The girl is beaming, wearing a T-shirt, incidentally, that announces across the chest I LOVE BOYS. The mom is beaming, too, not because she condones the message on her daughter's T-shirt but because "I have a new daughter." She hadn't had a single seizure since surgery.

According to the mother, the patient had now become the "older sister"—the more calm, conscientious, and mature one—even to the point of looking after her now "younger sister." She had been transformed. Her outcome was more impressive than I would have ever expected. While I was confident that the surgery would control her seizures, I didn't expect any miraculous personal transformation.

"Why do you think . . . ?" I asked my mentor.

"No more seizures. She's getting a full night's sleep every night," he answered.

At first I was incredulous—the explanation was too simple—but then I realized that it all made sense. If you mess with a kid's sleep every night, year after year, she's going to become a certain type of person. The disease will actually mold her. If you fix the disease, then the true person, submerged for so long in a fog of seizures and chronic sleep deprivation, can finally blossom into her true self.

I wish all the stories from my fellowship year could have been so happy.

My husband and I joke about the "old sneaker phenomenon." When you wear a pair of sneakers over a long period of time, they develop a worn-out appearance, and that worn-out appearance, at some point, reaches a transition point between being acceptably worn out and unacceptably sloppy. When you're in college, sloppy is fine and maybe even desired. When you get older, though, sloppy is usually just embarrassing.

The funny thing is: the person wearing the sneakers becomes too familiar with them, and cannot always detect when that transition point has been crossed. That's one advantage of being married. We're always honest about alerting each other as to when it's time to throw out "an old pair of sneakers." Neither of us takes it personally. We understand our own susceptibility to this universal human phenomenon.

In the epilepsy surgery clinic we carried around a black bag full of educational models, including a detailed life-sized brain that we could show to patients. The other material—sample electrodes that we implant over the brain and a sample vagus nerve stimulator—was stored in individual plastic bags. These bags would have been fine except for two things. One, they were biohazard bags, marked with an orange "biohazard" sign, of the type meant for transporting possibly infectious material like blood and urine samples. And two, they had become all crumpled up and worn out from sitting at the bottom of the large black bag, often squashed by the plastic brain model.

If I were a patient, I might be a little suspicious of anything that a surgeon took out of a crumpled-up biohazard bag, so I eventually took it upon myself to replace the bags with new ones. I suspected that the old sneaker phenomenon had been at play for a while.

This fresh switch, though, came too late for one particularly sensitive patient. She was pleasant, intelligent, and overwhelmingly in touch with all things holistic, alternative, and organic. She had numerous dietary issues, unrelated to any true allergies per se, but more related to various beliefs and psychosocial sensitivities. She insisted on bringing her own small refrigerator into the hospital. She would eat only her own food.

Her epilepsy came about after a neglected infection had affected her brain. She had been hesitant to seek traditional medical care despite several days of an unusually severe headache. Now, several years later, after her reluctant trial of numerous antiseizure medications, she would undergo an evaluation to see if we could localize her seizure focus and, potentially, offer her a surgical solution.

Unfortunately, her case was not straightforward and we were going to have to implant a series of electrodes over the surface of her brain—"invasive monitoring"—if we were to have any chance of helping her. She had a number of questions for me regarding the electrodes, so I fetched the black bag to facilitate my answers.

That's when I had to reach for the crumpled biohazard bag. I tried to shield the "biohazard" but it was so large and so orange. She recoiled. I pulled out a thin electrode strip, made up of a series of small platinum discs embedded in a silicone material. The wire trailing from it was a bit kinked in spots along its length. I explained that we would be implanting a series of these in order to capture a detailed recording of her seizures.

"So . . . these aren't natural . . . are they?" she asked me, already knowing and fearing my answer. I told her what they were made of. Her brow furrowed. I wished I could have offered her an organic soy and rice paper version.

"Honestly," I told her, "everything is natural, if you think about it. Are there really any unnatural things? Everything has to originate from something natural, something from the earth." My brief, pathetic soliloquy was completely unconvincing, but she ended up consenting to the placement of the electrodes anyway, even more fed up with her seizures than with the tragic state of the inorganic world.

Her frailties and hypersensitivities aside, I really got to like her. She maintained a cautious optimism. With the amount of controlled torture we were about to put her through, I couldn't bear the thought that there was a chance we might not be able to help her. Over the next several days, she endured the surreal experience of implantation surgery and then the epilepsy monitoring unit, with twenty-four-hour cameras watching her behavior, dozens of wires exiting her shaved monk head, and a parade of specialists tracking in and out, day and night.

In the end, though, we were forced to admit to her that we couldn't pinpoint the focus of her seizures. We couldn't offer her a surgical solution. We had put her through all of that and had come up with nothing. We took her back to the OR, took out the electrodes, and closed everything back up, defeated. I looked down at her head before wrapping it in white gauze. It was going to take a long time for her hair to grow back. At that moment, I wished she had fled from that crumpled biohazard bag, back to her organic world, never to be seen again.

Some people are lucky enough to be able to spend their entire career—their entire adult life even—in a singular pursuit of a singular focus. They have such a passion for just one specific thing, one disease, one treatment, one musical instrument, one social ill, that they are content to leave all other possibilities aside. This is a good thing. A life with a singular focus is a setup for excellence.

What I am passionate about, more than almost anything else, I'm afraid, is learning something new. If I could somehow benefit society and make a good living as a professional student, crossing disciplines as I went, I'd be tempted—very tempted. I love life on the learning curve and my curiosity is broad. This can be a double-edged sword, though, when it comes to settling on a career. Although it's true that in any given specialty, no matter how narrow the focus, you're "always learning something new," as every good mentor says, my suspicion was that I would always want to find out what, exactly, that new something was in the next conference room, operating room, or country on the other side of the world.

I would have to obsess over these abstractions later. This particularly cerebral year in my training had ended, and there was one more to go, but one that would not allow for quite as much time to think.

SEVENTEEN

Chief Concerns

The seventh and final year of neurosurgery training is the "chief year," when a resident is finally dubbed "chief resident." This year comes not only with a new title but also with new responsibilities. For one, you're in charge of trying to keep the peace among the other residents on the team. (All young and male, in my case. I've joked that I should have written a book entitled *Gorillas in My Midst.*) You don't want little problems to become big problems. You need to solve them before they grow large enough for the chairman of the department to take notice. In addition to your role as peacekeeper, you run morning and evening rounds, delegate items on the daily to-do list, try to keep all the attendings happy, and get everyone to the OR on time, by 7:30 a.m.

This year does have its perks. You get to choose the best cases. The attendings allow you more free rein in the OR. You tend to be shielded from minutiae because the other residents take care of much of the grunt work. You finally settle on some of your own opinions—how're you're going to handle a particular type of case, what approach you will prefer to take, and how you will speak with a family. You cement your own style. For all these reasons, the chief year is memorable for any neurosurgeon. In my case, though, there was one other factor that will always mark it as hard to forget.

My year as chief resident was the year of the World Trade Center attacks. I was in the operating room at the time. I was performing a craniotomy with a foreign-born neurosurgeon who happened to be from the Middle East. We were about to remove a cystic tumor from a patient's brain, and our world was about to change.

The technician who was monitoring the patient's brain waves for us had gone out for a break and was in the OR lounge, while the junior tech covered the monitors. We had a connection to the outside world.

"A plane just hit one of the Twin Towers!" our tech announced on his return to the OR. This set off a buzz around the room: What was going on, how could such a thing happen, who among us knew someone in New York? He peeked over his colleague's shoulder at the computer screen and went right back out to gather more information. We kept working.

Not too long afterward, the OR door swung open again, with a second announcement: "The other tower was hit!" We had opened the dura by that point and were entering the brain. We said nothing for a few moments while we continued to work. My attending was the first to speak and said to me quietly, without shifting his gaze away from the task at hand: "This is terrorism."

In my relative innocence, I had thought only "freak accident" during those silent seconds. My mentor, more realistic and wise to the world, knew better. He had already figured it out. Once he introduced the word, of course, it seemed obvious. It had to be terrorism.

Never before in my memory had current events entered the isolated cocoon of the OR with such urgency. A few years earlier, I remember listening to a live radio program during a case. Everyone wanted to catch the O. J. Simpson verdict in real time. There was a collective gasp in the room at the pronouncement of innocence, and then back to work. Other than that (and, on occasion, the announcement of important sports scores by an enthusiastic anesthesiologist), the feeling is usually that the news can wait.

(By the way, there is no need to worry about the patient's safety here. There are key moments of concentration during certain cases when everyone in the room knows not to talk—such as when an aneurysm at the base of the brain has just been isolated and is about to be clipped—but surgery does not otherwise demand absolute silence, and discussion does not have to be limited to immediate surgical concerns. The OR can be a lively social environment, unless the surgeon is in a rotten mood.)

Later, we would find out that another plane was missing somewhere in airspace not too far from us, and I even heard rumors that our hospital was on alert in case we were needed. The tragedy, of course, was that we weren't needed. Regardless, at least one of our neurosurgeons postponed his elective afternoon cases until later in the week, not because he was told to, but because he felt he just had to.

Once we finished the critical part of the operation and began the routine of the layer-by-layer closure, I let my mind wander. My husband was working in Connecticut, just outside of New York City. For career purposes, we needed to spend this one year apart, in different cities, seeing each other only on weekends. I was sure there was no reason for him to be in New York that day, but I would call him from the recovery room, just to make sure, as soon as we wheeled our patient out there.

He was in his office in Connecticut and watching the news with business colleagues from Seattle who would soon realize they were stranded. I thought about my brother. He was a starving artist in

Brooklyn, living an unstructured bohemian life. The World Trade Center was the last place I'd expect him to be, but I called my mother from the recovery room right after talking to my husband. She told me that my brother was in Manhattan but not downtown, and was making his way back to Brooklyn, on foot.

Then the ICU paged me, so I left the recovery room and walked over there. Televisions were on in every room, flashing unbelievable images at the bedside of each comatose patient. Everyone in the ICU who was conscious—nurses, critical care fellows, respiratory techs—tried to catch bits and pieces of the news as they went about their routines. I attended to whatever minor crisis it was that required attending to and pulled a chair up to one of the televisions, watching and listening over the sounds of the ventilator and heart monitor attached to the patient next to me.

I didn't linger for long. As the chief, I wanted to check on what the rest of the team was up to during this strange time. I walked over to the office and found half the department in the conference room watching the large screen that we used for our multimedia presentations. We had television access through this system, so the conference room had been transformed into CNN central, with my colleagues walking in and out as their pagers went off, vying for their attention.

There are isolated moments, turning points, in a life when a certain loss of innocence is realized. As the oldest of four kids in my family, I can remember feeling conflicted at times when I realized that one of my younger siblings was no longer just an innocent little kid. It's not that I didn't want them to grow up, wise to the reality of a sometimes harsh world, but more that I knew I was going to have to change the way I thought about them, and that's what was difficult.

Before I started medical school I even had fleeting thoughts about the fate of my own happy-go-lucky innocence. Did I really want to expose myself to suffering, trauma, and death as a routine part of my job? I've always been known as a generally upbeat and happy person.

What would several years of surgical training do to me? Would I be the same person and, if not, would that be for the better or worse?

In deciding to become a surgeon I promised myself one thing: I would never become the bitter female surgeon. Better to make a radical career change, if necessary, than to lead a stable but bitter life. It's not that bitterness is uniquely female in any sense. Believe me, I know plenty of unhappy male physicians. It's just that I had come across a couple bitter female surgeons along the way and they scared me a lot more than the bitter male ones did.

It could have been that the selective few I had met were older and had gone through their training during a different era, when the idea of women in surgery was still relatively new and not wholly accepted. Still, I resolved never to let myself fall down the slippery slope of harsh behavior, easy profanity, and loss of femininity that I imagined might lurk in the dark corners of a male-dominated specialty.

Now, after four years of medical school and into the seventh and final year of my neurosurgery residency, I was sure that I hadn't been irrevocably tarnished in any way. Plus, I had gotten to know plenty of other young female surgeons who were reasonably content, women who I knew would be excellent role models, the type of role models that female surgeons of the previous generation may not have had.

Certainly, a few distinct things did embitter me along the way—taking care of drunk drivers for one—but I had otherwise come through largely intact. I wasn't a totally different person, just more knowledgeable and experienced. One of my colleagues even mentioned that, whereas he had limped across the finish line, I was finishing strongly.

Still, I did have lingering concerns looking forward, concerns that I knew others shared and also sometimes repressed. Would I really want to do this for the rest of my life, the same tragedies, the same procedures, over and over again? What about being tied to a pager? I can tolerate the three a.m. call to a trauma now, but would I continue to tolerate getting out of bed for a drunk driver at age fifty? What about the need to plan my life around a call schedule? How would things change if and when we added kids to the mix? And perhaps worst of all—how bitter would I become with my first lawsuit? My trusted friends out in practice warned: "You *will* be sued, no matter how good, careful, or thoughtful you are." Nice, I thought, something to look forward to.

I remember being in a pediatrician's office as a kid and checking out a piece of artwork hanging on the wall in the waiting room. It was a cartoon type of drawing of a female pediatrician at work with chaos all around: children running around and screaming, tugging at her white coat, overstuffed charts falling off of shelves. The doctor was portrayed as positively harried, with slightly unkempt hair, glasses a bit askew, and a weak, tentative smile. The drawing was personalized, with the pediatrician's name on her white coat pocket. I imagined it was probably a gift, from a family member or a grateful parent. I think it was supposed to be endearing—a very human portrait of a busy working woman trying to keep it all together—but it was unsettling to me, even as a kid.

I couldn't fully intellectualize this gut feeling of mine until I grew up. Here's what I figured out I must have felt: Why would a woman intentionally choose to portray herself to the world as frazzled? Even if she does feel overworked and frantic from time to time, why would she want that image hanging on the wall? Why would someone create that kind of portrait as a gift? Maybe there is a comforting solidarity in expressing one's frazzled nature (assuming that every other woman must be in the same boat), but that's a little sad. Why glorify it? It reminds me of when smart women intentionally act dumb. I, for one, don't find that cute. Does the male equivalent of the frazzled doctor cartoon exist? I doubt it, but if it does, you'd probably never see it hanging proudly, as art, in his office waiting room.

One of the responsibilities of the chief resident is to run M&M. Morbidity and mortality conference (the official name, rarely used) is where complications, deaths, and other such problems are aired, formally, in an educational and sometimes contentious environment. It can be fun. M&M conference is when we get to see our mentors butt heads and our fellow residents get "pimped" (asked difficult questions) on a weekly basis. Along with the responsibility of running the show comes a small amount of power. Although others might recommend which cases should or need to be presented, the chief resident usually makes the final call.

Years before I had become chief, one of our attendings was fired and the residents always wondered if they had partly fueled the decision by weighing down M&M with his cases. Every neurosurgeon has vexing cases, complications, and patients who end up dying for various reasons—especially in a big academic institution that receives the most high-risk cases—but we only have time to present two or three patients each week. Only some cases get the spotlight and the third degree.

Most of the morbidity and mortality was due to the patient's own disease process, trauma, or just plain "poor protoplasm" (an unhealthy body with a natural tendency toward disaster), but actual mistakes were made on occasion. These were aired and discussed with particular gravity. Different surgeons have different styles in discussing their own mistakes, from spreading the blame to humble admission of singular guilt.

I will always remember how impressed I was as a young medical student in the audience of a neurosurgery M&M conference, listening to the chief resident at the time admit a technical error of his own. He was well respected by the staff and fellow residents for his combination of skill, judgment, and intelligence. During one particular operation, in an unlucky moment, he veered just a little too far to one side in dissecting the tissues at the junction between the back of the head and the top of the neck, an area where the vertebral artery—one of the critical blood vessels that supplies the brain—is located. There is some variability in the position of this vessel dictated by its tortuosity and the exact path that it takes in any given individual patient. Getting into this vessel is a known but uncommon risk of certain procedures in that area.

The chief resident on the hot seat at that moment, seven years my senior, stood at the podium and discussed the case, very matter-of-fact. He discussed his technique in detail, how he ventured out laterally, and how he injured the vessel, ultimately having to sacrifice it altogether to control the profuse bleeding that ensued. Luckily, the patient had enough overlap in circulation from the paired vessel on the other side that a major stroke did not occur, but that was pure luck. The same complication could have had a much different outcome in a different patient.

What impressed me was this guy's confidence and honesty. He wasn't the type to make careless mistakes, but mistakes sometimes happen, even to the best. He was appropriately apologetic in a professional way, without groveling, and did nothing to diffuse the blame or come up with convoluted explanations as to why it couldn't have been avoided. The funny thing was: afterward, there was no discussion, no questions, no contentious accusations. There was no need, and he continued on to the next case.

Seven years later, it was my turn to run the show, this form of medical theater and internal policing. Our M&M conference adhered to a time-honored ritual. The chief resident, of course, stood up front at the podium. The department chairman sat in the front row, in a corner seat, with the black three-ring binder that held a list of all the neurosurgery operations performed that week, including who was involved in each case and whether or not there was a complication. There was otherwise no formal assigned seating, but the residents tended to favor the back of the room (the post-call resident might try for the last row, allowing him to rest his head against the wall).

Logic would dictate that if you hid yourself well enough in the room you wouldn't be "pimped," but logic was proved wrong again and again as the chairman would crane his neck and body into extreme contortions to make sure he saw the guys in the back who were trying to hide. They were sure to get grilled. Still, exposing your flank, right up front in the open, seemed even riskier, so we all persisted in playing hide and seek.

Here's how a presentation would go. The chief—me, in this case—would present the first patient in the formal medical parlance and format, the same exact style used by generations of surgeons before us:

"L.M. is an eighty-five-year-old woman with early dementia who complained of a headache and then developed a dense right hemiparesis, confusion, and aphasia." (In other words, she couldn't move the right side of her body and couldn't speak.)

At this point the chairman might jump in with a question, usually for the youngest resident on the team. He was good at asking age-appropriate questions, pimping the intern or junior resident on basic anatomy, physiology, and pathology questions and reserving detailed inquiries regarding the pros and cons of various surgical approaches for the senior or chief residents.

"John: What do you think is going on here?" the chairman would ask.

"Uh . . . could be a bleed, stroke, tumor . . . infection," John might answer.

"Infection? Come on now. You don't really think this is an infection. Why do you guys always feel obligated to give us the most comprehensive list? Tell me what you really think this is."

"A bleed, sir."

"Good, go on."

(It's hard for a resident to win this game. If John had simply said "a bleed" with bold confidence from the start, the chairman very well could have answered "Is that it?" and chided him for neglecting to list other possibilities.)

The chairman would leave John alone for a minute and would turn his attention back to me as a sign to continue with the case.

"Her past medical history was significant for hypertension, peripheral vascular disease, depression, and heavy smoking. She was on aspirin, Prozac, labetalol, and Lasix. She presented to the ER where a stat head CT was performed, and we were consulted."

"All right, back up," the chairman would say. "What about her exam?"

"Okay," I'd continue, "in the ER she was awake but sleepy, nearly flaccid in the right arm and leg, unable to follow commands and unable to speak."

"Pupils?" he'd ask.

"Equal and reactive."

The residents always wanted to get to the visuals as quickly as possible: the scan of the brain, the money shot that would give us something to work with. The attendings would annoy and taunt us by questioning us on details of the neurological exam in an effort to teach us a thing or two, or to criticize us for failing to check for a specific esoteric finding. In reality, we usually saw the scan before we saw the patient, but we would make an effort to follow the rules of formal medical presentation anyway, with history and exam first, then scan, albeit artificial in the modern era. After a few nitpicky questions about the exam, followed by my perfunctory answers, and maybe another question lobbed over to John about what the scan might show, the chairman would give in:

"Okay. Show us the scan."

I would flash her head CT scan up on the screen, cut by cut, from the bottom of her head to the top. Then I'd take the initiative to pick on a resident myself and ask him to take us through the findings. This is how the conference would go; a step-by-step unveiling of the story punctuated by pointed questions to the audience. The best part would be the debate that followed.

I liked to request: "A show of hands for those who would recommend taking her to the operating room." Maybe a third of the room would be in favor. "And who would *not* operate?" The majority was in this camp, but a few didn't vote, so I would single out one of the unlucky few who couldn't decide.

"Mike, you didn't vote. What are you going to do here? You have to make up your mind."

Mike was one of the younger guys, unsure of the right decision here.

"Well, she's eighty-five, a big smoker, not the healthiest, she's probably not going to do well. You might not be able to get her off the ventilator. But . . . the bleed . . . it's huge. She'll probably end up dying if you don't take her to the OR. Maybe you just go in and take the clot out, hope for the best, but——"

"Mike, make up your mind," one of the senior attendings pipes up. "You're on call. You're in the ER. What's your recommendation over the phone to your attending?"

"Okay, okay," Mike answers, a little flustered, "I'm taking her to the OR."

"So, you're taking an eighty-five-year-old hypertensive smoker to the OR?" one of the pediatric neurosurgeons asks. "I thought we all knew the data on intracerebral hemorrhage in the elderly. You might improve her short-term outlook, get her off to a nursing home in a few weeks, but she'll end up doing very poorly in the long run, just the same. You're just prolonging the inevitable."

"Wait a minute!" One of the vascular neurosurgeons jumps in. "I've had plenty of families who have thanked me for taking their loved ones to the operating room in this type of situation. Patients don't always follow the rules. Some can do quite well. Come to my clinic, you can meet a few."

"Your idea of 'quite well' is in a nursing home with a feeding tube!" the pediatric attending says. The room erupts in laughter.

The chairman tries to maintain some order. "Okay, that's enough. Let's hear what happened to this poor lady."

I continue: "The recommendation was made to send her to the ICU, stabilize her, but not to operate. The entire family was okay with this plan except for the daughter, the one from out of town, of course. She asked for a second opinion and had heard of Dr. Jones through a friend of a friend. Dr. Jones saw her the next morning and said that surgery was an option and that he would actually be in favor of it. The rest of the family finally gave in to the daughter, and the patient ended up in the OR within twenty-four hours of admission."

"Hmm . . . sounds a little complicated. How'd she do?" the chairman asks.

"Well, she woke up after surgery, more alert, and was starting to move her right side within three days. We couldn't get her weaned off the ventilator because of her lungs, and she ended up with a pneumonia."

"Big surprise there!" I heard one of the more cynical senior residents mutter under his breath.

I continued. "Finally, after serious debate, the family decided on the trach and PEG route (tracheostomy and feeding tube surgically inserted through the stomach wall), and now we're awaiting nursing home placement."

"Nice case," concludes the pediatric neurosurgeon. "Next?"

In a department made up of over twenty neurosurgeons, as was the case at my training program—one of the largest in the country— you will see the entire gamut of clinical decision making, from the most conservative to the most aggressive. As a chief about to go out into the real world, I felt lucky to have been able to sift through the large collective experience of all of my mentors and develop my own set of philosophies and leanings that would inform my recommendations going forward. But that was the scary part of anticipating the end of the chief year: soon, the buck was going to stop with me. No longer would I be able to suggest a strategy to my attending over the phone at three a.m., discuss the pros and cons, and then have him look over my shoulder as I carried out the plan that he approved.

The interesting thing is, different residents, drawing from that same collective experience, may come up with different views of the vast array of neurosurgical options by the end of it all. Medicine is a human endeavor. Here's a secret, then: what happens to you on admission to the ER for a neurosurgical problem may not be based purely (or even mainly) on science. The science regarding a particular situation may well be scant or conflicting. The individual philosophy of the neurosurgeon who happens to be on call that day may be of equal influence. As evidence, M&M conference is chock full of debate, often the same debates over and over again, back and forth, especially the "operate versus don't operate" variety.

The burgeoning elderly population—prone to falls, catastrophic bleeds from blood thinners or worn-out vessels, and brain metastases from recent or remote cancer—ensures, unfortunately, that there will be no shortage of cases to be debated. (Over the course of a two-day period recently, I was asked to consult on three patients in the hospital. Their ages: eighty-five, ninety-three, and ninety-nine years old. All involved some form of blood in the head.)

You can't always state a treatment philosophy with absolute certitude (except maybe the one I learned in pediatric neurosurgery: if the mother thinks her child isn't quite right, then the child isn't quite right, even if he looks perfectly normal to you). There are so many variables to consider in difficult cases, especially in the older patient: Is the person a frail eighty-five or a hearty eighty-five? What is their baseline quality of life like? Are they demented, and to what degree? What preferences had they expressed ahead of time? What are their family's wishes? How complicated are the possible interventions? How much "torture" are we willing to put a person through and for what possible outcomes?

It can all boil down to this question: What constitutes a life worth living? This is where it gets messy, philosophical, and personal. Consider life in a nursing home at age eighty-five, unable to speak or move one side of your body, and unable to care for yourself. One colleague of mine told me that as long as he could hold his grandchildren on his lap, then that would be a life worth living. Another responded, when presented with the same scenario: "No way! Spare my family the anguish." Our recommendations to others can very well be colored by what we feel would be an acceptable life for ourselves.

How much value do statistics play in these sticky decisions? What if one study concludes that 10 percent of elderly patients operated on for large bleeds have a "good" outcome, broadly defined. What if another study says it's only 1 percent? What if you think those studies, probably uncontrolled and retrospective, are no good to begin with (as many studies are)? What about the one-in-a-million miracle case from the news or the tabloids that families always ask about, the person who came out of a vegetative state after ten years, still bed bound but able to smile and utter a few words like "hi" and "Pepsi"? Those stories make our jobs more difficult.

Once a decision is made to take someone to the OR, when the situation is not black and white but gray, and when the ultimate outcome after surgery remains uncertain, you almost have to say "we did the right thing" no matter what. If not, you risk casting doubt on the decision, inciting feelings of guilt, and deflating everyone's morale.

I remember taking care of a patient who traveled from several states away for consultation regarding a very rare and vexing problem: intractable hiccups. This was no joke. The patient was quite elderly but otherwise reasonably healthy. His hiccups started soon after coronary artery bypass surgery a few years earlier, and they were ruining his life. He had already tried everything and had had every scan of every possibly implicated body part. His family did the research and found out that one of the star neurosurgeons in our department had a reputation for curing unusual and often mysterious ills that may have their origin at the brain stem, and a couple cases of brain surgery for intractable hiccups had actually been written up as case reports in the medical literature.

We don't see this kind of thing every day, or even every year. No one does. Despite the uncertainty over whether or not surgery would help, that's what the neurosurgeon recommended to this elderly gentleman, who quickly consented to the procedure, with very little to go on except for a couple previous cases worldwide and the confidence of a single senior surgeon. He was desperate. And, truth be told, the neurosurgeon in question was one that I would put in the genius category, albeit of the mad scientist variety, and if anyone could help this patient, this surgeon could.

The patient had a rough recovery. The operation had sapped his strength. His hiccups did improve—somewhat—but they did not go away, at least not during the time I knew him, while he was still in the hospital. Before sending him back home, across the country, his neurosurgeon hoped the patient would tell his story at the weekly teaching conference that we shared with our neurology and neuropathology colleagues. He would be a great teaching case, and he presented the perfect opportunity to highlight a complex and delicate operation. The patient agreed, and I was chosen to present his story and treatment in the usual format, with the patient up in front of the room facing the audience and the attending neurosurgeon seated in the front row, available for comments and questions from the audience.

The only wrinkle in our plan was that the patient looked and felt as if he had just been put through the wringer. He was elderly and hadn't yet bounced back after this major ordeal. He had already stayed in the hospital longer than we had predicted. He was still too weak to walk long distances or to stand for any length of time. I would need to transport him to the conference in a wheelchair.

The neurosurgeon had arrived at the conference room a few minutes early, and I showed up not too long afterward, wheeling the patient along. Upon our arrival, my attending jumped out of his seat and rushed over to us. He whispered to me: "We can't have him sitting in a wheelchair like that in front of everybody. Let's have him stand."

"But he'll have to stand for half an hour and——" I protested.

"He'll be okay," my mentor replied.

So I led the patient, arm in arm, to the front of the room and hid the wheelchair somewhere down the hallway, before most people had shown up. I had the patient stand right in front of a desk at the front of the room so that he could lean on it if he needed to.

I got started with the presentation in front of a full audience, breaking at one point to have the patient recount his own hiccup saga. His voice was a bit weak. I watched him closely. He was starting to lean more and more against the desk, first one arm and then two, and I worried he might collapse if I didn't finish the presentation quickly.

I rushed through the rest of my talk, going over the history of hiccup treatments and the rationale behind a somewhat controversial operation. The patient was now trembling, exhausted, and trying his hardest to suppress his intermittent hiccups in front of the audience and his surgeon.

I ended the talk a little early and ushered our patient out of the room and back into the wheelchair, feeling as if I had just been an accomplice to something a bit subversive, something that felt strange. I don't know, in the end, if we really did the right thing for this kind gentleman, but I certainly knew what to say, encouraging him as I wheeled him back to his room: we did the right thing.

———

I'm thirty-three years old when my chief year ends. *Thirty-three.* I've been in some form of education or training up until this point: college from age eighteen to twenty-two, medical school from twenty-two to twenty-six, neurosurgery residency from twenty-six to thirty-three. My friends who aren't in medicine have been out in the workplace for years already. I hear about CEOs of Silicon Valley start-ups who are younger than I am. I read the occasional news story about a popular new mayor—my age—elected to office. Hollywood actresses in their thirties are nearing the ends of their careers, which is okay, because they've already earned millions. Many professional athletes, at the apex of their professional life, are younger than I am. My parents had a houseful of kids by the time they were my age. I feel as if I have some catching up to do.

I recall the warning we all received when interviewing for our seven-year apprenticeship: "Don't go into neurosurgery unless there's *absolutely nothing else* you could ever see yourself doing." This warning should also come with the reminder that "You won't emerge from this tunnel until you're in your thirties."

I hear about other professionals "reinventing" themselves from time to time, trying out a different career, pursuing their intellectual curiosity to various ends, or taking a few months off between jobs to, say, trek in the Himalayas. That type of thing doesn't go over so well in medicine—particularly surgery—which is often seen as a sort of "calling" and not as amenable to professional wanderlust. Here's what people would say to a surgeon who wanted to take a look around: "After all those *years?*"

But we reassure ourselves by recalling all the things we've experienced that our white-shoe friends in business haven't, the things that get our hands and shoes dirty and that make a difference in people's lives. We can always regale in those highs, and look forward to more. At the same time, though, it's easy to be haunted by the very experiences that most starkly set our job apart.

EIGHTEEN

Endings

Looking back at seven years, I have had my hand in saving lives and I have had my hand in helping to end them. I'm not talking about murder, of course. I'm talking about helping people die; people who already have their toes at death's door but are about to cross the threshold in a very unpleasant and—to use a catchword—very undignified way, the type of way that would burn a horrifying imprint into the family's collective memory. Some might argue that this is never a doctor's duty, but I would beg them to take another look, with their eyes open.

Take the eighty-seven-year-old woman who collapsed in her kitchen while spending a pleasant day with her granddaughter. She arrived via ambulance at our ER in a coma but was still able to breathe on her own, somewhat. Her family showed up a short while later. With no advanced directives to go on, she was intubated by an

ER physician, put on a ventilator to assist her breathing, and sent for a scan. She had suffered a massive spontaneous intracranial hemorrhage. Our best guess, based on her age and the location of the blood, was that she had had an "amyloid bleed," a type of bleed particular to an elderly brain with worn-out fragile blood vessels that are prone to breaking for no particular reason. No trauma is required and it's not well understood. The dreaded "why" question is not fruitful here.

I met the family in the ICU waiting room. I explained everything: the scan, her exam, her dismal prognosis. I had already spoken to the attending in charge. He was in favor of no intervention. We wouldn't even offer the option of surgery. He had taken a peek at the scan and the patient and then left me on my own to do the right thing.

One of the daughters was a veterinarian. I showed her the scan, holding it up to the fluorescent overhead lights of the waiting room. That's all she needed. She told me that her mother wouldn't want to live like this. She had spent a great day with her granddaughter. We should end things on a happy note. No one disagreed.

I led the family down the hallway to the end of the ICU where the patient was lying immobile, comatose, on the ventilator. I left them alone and told the nurse to page me when the family was ready. My pager went off soon afterward.

I returned to the ICU and took the veterinarian daughter aside in the hallway. I wanted to run something by her. The area of hemorrhage was quite high in her brain, and her brain was markedly atrophied, given her age. In other words, there was a lot of room in there to accommodate all the blood. For those reasons, there was no real pressure on her brain stem, the part of the brain that controlled breathing. She might very well breathe on her own after being detached from the ventilator, maybe for hours, maybe for days or even longer in this unconscious limbo.

I had seen such things before, and it can be horrifying: an elderly patient sent to the floor for what is thought to be their final hours, and the hours turn into days. The patient gasps and sputters with irregular breathing (giving the appearance of distress), foul odors collect in their mouth and throughout the room, a pneumonia and a bad urinary tract infection set in, no one wants to hold vigil by the bedside, family members visit less frequently, and when they do they prefer to mill around in the hallway. I wasn't going to let that happen. I swear, sometimes we treat our dying pets more humanely.

I told the daughter that, with the family's permission, I would order a morphine drip, to be used as needed. She understood perfectly, and thanked me profusely, far more than most families do, even after a successful save, after several hours in surgery.

I spoke to the nurse, found the patient's chart in the rack at the nurses' station, and scribbled the final orders: "1. extubate patient. 2. MSO_4 gtt titrate prn for respiratory distress. 3. CMO." The "comfort measures only" order made things official, ensuring that no other well-meaning members of the team would come by and complicate things by ordering lab tests, medications, or follow-up scans.

The "titrate" addition to the order for the morphine drip meant that we had the flexibility to escalate the dose as we saw fit based on her breathing pattern, knowing that progressively higher doses would mean progressively slower and shallower breaths. In this way, the dying process could be curtailed and made more humane for both the patient and the family. (More humane, I think, than the "passive euthanasia" of withdrawing food and water. Most people wouldn't consider starving the family dog as a way of "putting it out of its misery.")

Shortly after extubation, with the clear drip running through her IV, our patient's life ended—calmly—with her family encircling her bed, with fresh memories in their minds of the woman she had been just hours before.

The first gift I ever received from a patient or, actually, from a patient's family was as a junior resident working at the Veterans Administration hospital. It arrived wrapped very neatly in brown paper. I had never received any sort of personal package at any hospital before. I was only a transient worker bee at the VA, doing time on my required four-month rotation, and I was a bit surprised that the package had actually found its way to me through the unwieldy government system of the hospital.

The gift was a Whitman's Sampler box of chocolates. The thoughtful gift giver was the wife of a former patient of mine, a patient of very humble means, a war veteran who had died while under my care. At the time of his hospitalization I had felt fairly impotent, unable to offer anything but a dignified death in the face of advanced metastatic cancer. This simple box of chocolates, of the finest variety to my new way of seeing things, made me realize that "nothing left to offer" wasn't quite true.

———

Sometimes, in dealing with a difficult reality, I am struck by how well art has imitated that reality, if not in every detail, then at least in overall feeling. A perfect evening for me is dinner and a good movie, and I especially love when a movie haunts me, provoking heated discussion with my husband at a café afterward, late at night. I'll always remember the raw, provocative, and morally ambiguous scene in a Spike Lee film from several years ago that really got us talking. One of the main characters is a minister. His son has gone bad, having fallen prey to the primitive biological forces of drug addiction, and now he's past the point of no return, the chemicals and his need for them having taken over his brain with a force too powerful to be defeated. In his unbridled quest for money to buy drugs, he has stooped so low as to terrorize his parents in their own home, and his parents now fear for their lives, and rightly so.

It's almost as if a rabid animal with a virus-laden brain had been set loose. The fear and violence escalate to the point where the parents are forced to defend themselves or risk death, and the minister ends up shooting his own son, ending both his son's life and the tragedy of the end-stage drug abuse that had threatened so much else.

I once admitted a patient who I thought, by the family's account, could have been the inspiration for the character of the minister's son. He was in his fifties, which, after meeting him, I felt was amazing in and of itself. From the looks of him I would have expected him to self-destruct much earlier. He looked at least seventy. He had had a very hard life of alcohol, cigarettes, and cocaine and he had failed all attempts at rehabilitation. He had done time, a few times. He had been violent. His toxicology screen on admission was positive for cocaine and alcohol, and the smell of alcohol emanating from him was so concentrated that it was nauseating.

He had fallen many other times before in similar drunken and drug-induced stupors, but never like this, from the top of the stairs. He tumbled all the way down and ended up at the bottom, quadriplegic, no movement from the neck down. He couldn't breathe, as his diaphragm had no input from the spinal cord.

The paramedics saved him on the scene, packaged him up, and handed him over to us, a gift. Now what?

The X rays of his neck told us all we had to know. He would be quadriplegic for life. The fracture and degree of dislocation at the top of his neck were impressive. We were surprised at how light and osteoporotic-looking his bones appeared on X ray, especially for his relatively young age. (Another reason, by the way, to avoid smoking at all costs—it's not just lung cancer, various other cancers, heart attack, and stroke that you have to worry about. Osteoporosis is also more common in smokers, and it only takes one broken bone to ruin your day, especially when it's in your neck.)

What were we going to do with this guy? You *always* want to give a patient the benefit of the doubt, if there's any hope at all. Maybe we'd be able to get him off the ventilator eventually, manage all his withdrawal symptoms, ward off the bed sores and infections that patients like this are prone to, and get him off to rehab, one day, where he would find religion and, via dictation to a dutiful nurse, write an inspirational book that would influence thousands and perhaps become the basis of a movie. Maybe not. His family certainly didn't think so and they knew him quite well.

His body didn't take well to the quadriplegia and we had difficulty controlling his dangerously low blood pressure. To begin with, his heart wasn't strong and his lungs were shot. What's more, a combination of low blood pressure and ratty carotid arteries (again, smoking) left his brain underperfused at times, his mind unavailable to us or his family. In his intermittent fully awake states, he could nod his head yes or no reliably. When I asked him whether or not he wanted us to keep him going in this state, he shook his head no, every time.

I spoke at length to the family, on multiple occasions, and they visited dutifully despite their mixed feelings toward him. He had burned many bridges over the years, but they were still family. His closest sister was a Baptist minister. After several days in the ICU, with his health failing further, she asked if we had to continue with all this care, leading nowhere.

In a case like this—a relatively young man intermittently awake and on a ventilator—I can't recall now if the ethics committee was involved in our collective decision making, but they probably were. Even so, they may have had nothing much to add. The decision to terminate life support was unanimous among everyone: the patient, the family, the physicians, the nurses. This was not the type of case to hit the media, and there were no demonstrators carrying signs outside the hospital in support of continuing aggressive (and free) care at all costs.

Before taking him off the ventilator, the family gathered in his room and sang church hymns, led by his sister. They left quietly and then we pulled the curtains and slipped the breathing tube out of his throat. His eyes remained closed while I stood at the head of the bed with his nurse, watching the heart and oxygen monitors. We waited several minutes. Despite the fact that he was not able to breathe, his heart kept beating, weakly, and his oxygen level dropped more slowly than I had expected. Then, without warning, his eyes shot open for a second—wide open—then closed again and that was it. I walked out to spend a moment with the family and, again, they thanked me with more warmth than I was used to.

His room now emptied out and disinfected, his family now back home, both sad and relieved, we accepted another patient into our filled-to-capacity ICU without skipping a beat, refocusing our energy and resources on this fresh new person who would have no knowledge of all the weight that had just occupied this very bed.

All it took was a single drop of blood to remember a patient I otherwise would have forgotten. It's easy to forget a patient whom you never really got to know, especially during residency when the pace is hectic, the volume tremendous, and the memories fleeting. In a really busy week, head injury victims might arrive daily, and one or two might not make it, but you have to move on.

Some victims arrive as "Doe" if the paramedics are unable to find ID. Family notification is delayed until someone can figure out who the patient is. We do whatever we need to do on an emergency basis—surgery, aggressive measures—without obtaining anyone's consent. In most cases we learn the patient's name later that day, but that doesn't necessarily mean that we get to know them any better.

I examined a Mr. Doe once in the trauma room within minutes of his arrival and knew he probably would not make it. He was a so-called multiple trauma victim: his brain wasn't the only thing injured. He had been unrestrained in the car—without a seat belt—and that meant that his body was catapulted forward at whatever speed the car was traveling at the moment of impact. (An air bag just isn't enough.) As I crouched down at the head of his bed to examine a bad scalp laceration, a drop of his blood landed on the right sleeve of my white coat. He died within twenty-four hours of arrival and I filled out his death certificate.

There was nothing unique about his case, nothing particularly memorable. I hadn't gotten to know his family aside from being the bearer of bad news. He hadn't been in the ICU long enough for them to bring in photos of him and hang them all around, or to display the traditional (and tragically insufficient, in cases like this) "get well soon" cards with personal messages that get me choked up. Those are the things that make me feel like I kind of know the person lying in the bed, unconscious, whom I have never truly met, and may never get to meet.

(One parent even hung a sign over her teenage son's bed reminding everyone of what radio station he liked to listen to—"I'm Nate. I like B98"—so that the radio in his room would be appropriately tuned to his individual preferences and not to the default soft rock he would have ridiculed before the accident, in his conscious state.)

I was so busy that week that I didn't even change my white coat. I just didn't have the minuscule amount of extra mental and physical energy required to: (a) realize that it would be most appropriate for me to change my dirty white coat, and (b) make my way to the closet in our office where all of our coats are hung after being laundered and rifle through them to find one with my name on it. It was that kind of week, and that particular task was the lowest on my list of priorities.

So I kept seeing this small, neat, round bloodstain in my peripheral vision as I went about my daily activities, and I kept thinking about this patient whom I otherwise would have quickly forgotten. Who was he? What was he like? Who were all the people who would be mourning him? What if he had simply worn his seat belt? What if his injuries had been less severe, and survivable? Could he have been one of the "great saves" that comes to visit the ICU months later, after rehab, and wows all the nurses?

If every patient left a stain, a resident's life could very well become an unbearable mess. A stain every once in a while, though, can probably help keep us human. That realization must have been one reason I kept a journal during my training. I knew I would otherwise forget along the way, partly because there was too much to remember and partly because I might want to forget.

NINETEEN

Confronting Age

My day-to-day is different now, out of neurosurgical training and into my first real job, so to speak. I work in a genteel Connecticut town outside of New York City. The hospital is well located. I could walk to the local Tiffany's after finishing a case if I ever decided that I wanted to go to a Tiffany's (lucky for my husband, I'm not that kind of girl). I could drive into New York for dinner, for the best Japanese cuisine outside of Japan (that's more like it). The cars in the patient parking garage are more upscale than in the doctors' parking garage. Many younger doctors—myself included—don't live in the same town as the hospital; it's too expensive. It's more suited for successful hedge fund managers and Fortune 500 CEOs, at least if you're looking for a home with a place to park and more than one bedroom.

The hospital always scores in the top 1 percent of the country in patient satisfaction. Patients have many reasons to be satisfied. The lobby looks more like a hotel lobby than a hospital lobby, lobster is available on the patient menu on certain days, new moms are offered champagne after delivery, and I'm told that a patient can request to have the art in his room changed if it doesn't please him. And, by the way, the medical care is excellent, too.

One small downside, I guess, is that the area is such a desirable place to live that many doctors desire to live there, or around there, along with everyone else, which means that it is well overserved. In other words, it is a competitive practice environment. I stopped once at the scene of a bad car accident in town, on a beautiful winding leafy road lined with estates. An SUV had slammed into an old stone wall on the property of a private country club, just seconds before. Two other cars had stopped, and so there were a couple other people on the scene already. Both were doctors. The victim had as many physicians at his disposal—right there on the side of the road—as he would if he had crashed directly into the local emergency room.

So the town is not like certain more remote parts of the country that are dying for a neurosurgeon and willing to offer huge salaries to lure one in. I get ads in the mail at least once a week with cryptic declarations that sound a bit desperate, like: "Practice only 100 miles from a moderate-sized metropolitan city!" My parents even get these letters for me at *their* home, although I haven't lived there for years. I have no idea how the recruiters even got the address. Once my mom called me and said: "You won't believe how much *this* one is willing to pay!" (But Mom, what kinds of restaurants would I spend all that money on out there?)

There may very well be more neurosurgeons available to the local population here than there are available to the entire population of sub-Saharan Africa. This is, I have to admit, a small source of guilt on occasion. Maybe I should go to a town that really needs me more. That said, I also have to admit that I thoroughly enjoy living here. One theme of my postresidency life has been of allowing (or catching up on) a touch of hedonism here and there, and for that I don't have any guilt. Still, I'd love to reach a point, some day, where I would do neurosurgery as volunteer work, for a more desperate population.

A few other things are different now, with my training in the past. As a resident, my youthful appearance wasn't much of an issue. Patients knew I was a resident—a young doctor training to become a neurosurgeon—and not their "main surgeon." Now that I am a fully fledged neurosurgeon, comments on my youthful appearance are common. Walking down the hallway in the hospital I might hear, "Hello, Dr. Firlik!" from a nurse who passes by. Then I may hear—usually from an older man who witnessed the exchange—something like: "You're not old enough to be a doctor!" I laugh, but I usually don't bother to say that I've been a doctor for going on ten years already.

I can understand when patients who are considering surgery ask my age, point blank. I've come to see it as natural, even for an otherwise polite and sophisticated clientele. It doesn't mean they won't trust me, it just means that I might have to work just a little bit harder to gain their trust than if I were, say, a man with gray hair or a bald pate (even of the same age). Regardless of what we're taught in elementary school, looks *do* matter. I saw a new patient recently who was referred to me by an already established patient of mine. Upon entering my office and seeing me for the first time she said: "My friend told me about you and recommended you highly, but she warned me to expect someone who looks like a teenager, so I was prepared. Nice to meet you."

It doesn't matter what kind of suit I wear and I haven't bothered to try short hair for a change, because I don't think it would make a difference. Sometimes I tell colleagues that I'd like to find a plastic surgeon who would be willing to create a few wrinkles for me. I don't foresee being able to tell that joke for too much longer, though, so I'll try to appreciate the inevitable age questions in the best possible light.

I have come to see my office as a forum for confronting age. I'm not talking just about my own age, but particularly that of my patients. An average day now revolves around confronting, or facing, up to the effects that aging and the wear and tear of normal life have had on their bodies, and then devising an appropriate treatment plan. This doesn't sound very glamorous. It may not be what you think of when you think about neurosurgery. Most neurosurgeons, however (except the selected ones in academia who specialize, for example, only in brain tumors), spend most of their time in an endless quest to treat a growing epidemic: the aging spine. The official term for this very common entity is "degenerative spine disease."

After I take a patient through the details of his or her scan, pointing out the areas of concern, I may get a comment like: "So what you're telling me is that I'm getting old," or "It's not fun getting old." I always keep a box of tissues at hand, just in case such a comment, or the fresh realization behind it, triggers tears. The fact that I get a sneak preview of "getting old" every day in my office is another reason I don't feel guilty about enjoying the here and now, while my joints remain pain-free. I don't expect it to last.

Arthritis refers to a degeneration of the joints. The spine—made up of the cervical, thoracic, and lumbar segments, as well as the sacrum—can be viewed as a long series of joints. Specifically, the discs (a type of joint) are located in the front of the spine, and the facet joints are in back. Because there is no cure for arthritis— only management—an important part of treating arthritis of the spine is managing expectations. Spine surgery can relieve pressure on a given nerve or nerves, caused by a buildup of arthritis, and so surgery can improve the debilitating pain shooting down a leg, but it cannot give a patient a brand-new spine. There is no such thing as a spine transplant, although I have had plenty of patients who have asked for one. Undergoing spine surgery or neck surgery does not mean that you will never have back pain or neck pain again. Aging is a progressive condition. Luckily, though, the management strategies that we have to choose from, both surgical and nonsurgical, do tend to work adequately for most people.

Younger people, oddly enough, can sometimes develop degenerative conditions of the neck and back, too, which always prompts the questions, vexing to both patient and physician: Why me and why so young? I have yet to come up with a satisfying answer, mainly because there really is no definitive answer, but sometimes I liken it to developing gray hair: some people get it earlier than others, but most people will get at least some of it sooner or later, especially if they live long enough.

One of my mentors likes to answer the "why me" question with "There are three possibilities: bad genes, bad habits, or bad luck." Degenerative spine disease may be a combination of all three, but I tend to downplay the habit part (unless someone is obese, which means the spine has been unduly stressed) because there's no use blaming the patient, especially when we can't pinpoint the habits that clearly contribute to the problem. A thin, healthy person with a desk job may develop just as degenerated a spine as an overweight construction worker. It's not worth spending too much time in the office trying to answer the why question, but there are some interesting theories floating around.

Some surgeons would mention that spinal degeneration is simply a by-product of living an upright life as a member of the Homo sapiens species, as opposed to our primate ancestors, but that doesn't answer the specific "why me" and "why now," because the vast majority of us lead upright lives most of the time. Plus, the unfortunate sequelae of evolution are hardly of interest to the average person whose back is killing him.

In an elderly person, I could also add that the human spine wasn't "designed" (although that's not the right word) to last eighty-plus years. And, because people aren't dying of all the infectious diseases they used to die of, they're living much longer and asking their spines to keep up the support, well past the point where their youthful, formerly well-hydrated discs have dried out. But that explanation does nothing to soften the blow either, or to answer a younger patient.

I recently saw a young man—late twenties—in my office who complained of annoying neck and shoulder pain for a few months, and had tried all the usual home remedies already. His MRI showed a degenerated disc in his neck. He had seen another surgeon who recommended a standard trial of physical therapy and anti-inflammatory medication, and warned that he may eventually need surgery, but that surgery might not be required for one, five, or ten years down the line. That surgeon also tossed in the concern that he could potentially be at "higher risk of paralysis" than the average person off the street if he were to injure his neck. The patient told me that he didn't like what the orthopedic surgeon told him and—even worse—how he said it. The out-of-network payment, he added, was the final straw. So he went looking for a second opinion.

I largely agreed with the other surgeon, in broad terms, but I explained the situation in a gentler manner. I also mentioned that paralysis, although technically possible, would be extremely unlikely, so he shouldn't go about his life harboring a new neurosis. As it is, a surprisingly high number of our patients check the "anxiety/depression" box on our intake survey of medical history. I would hate to contribute to that.

"Look," he said, "I just want you to tell me it's okay to play football. Okay? I'm a really active guy." He almost sounded a bit threatening.

Here's where my lawsuit detector goes wild, even though the patient was not thinking along those lines. The patient wanted to play football, wanted to lead a normal life, and he wanted an assurance of no risk at the same time. I hated to say it, but that's not how the body works and it's not how medicine works. How could I predict what would happen to his neck if he were tackled head on, again and again? There is no study out there, for example, that gathered one thousand football players, scanned all their necks in the MRI scanner, let them play football, and then checked to see if the ones who had signs of disc degeneration on their MRIs were the ones who ended up paralyzed somewhere down the line. Not only has such a study not been done, it probably never will be.

So now that I had seen his MRI and now that he had asked, I was pretty much obligated to recommend against football. Because what if, instead, I said it might be okay to play football and then he went out to play, got tackled, and became quadriplegic? Even though a very rare injury like that may have nothing to do with his preexisting disc degeneration—it could be due to a fracture or torn ligament in the neck—try to explain those fine distinctions to a jury. All they would see is a cavalier doctor who said it was fine to play, and a young man whose life was ruined. (At $106,000 per year, my malpractice premium is already high enough.) I have to err on the side of caution, even excessive caution.

I'm not that crazy about football anyway, so I didn't feel too bad cautioning against it. One of my most depressing moments of residency was having to tell a college football player that he would likely be paralyzed from the neck down, for life. His neck was fractured, and his spinal cord pretty much converted to pulp, during a routine football practice. I'm actually surprised this doesn't happen more often, in a sport that intentionally involves repeated head-on tackling.

By that point, the patient probably wished he hadn't gone for the MRI in the first place, leaving his disc degeneration out of sight and out of mind. His next question was obvious: If not football, then what else was out? Was skiing okay? Well, he could fall down at high speed and injure his neck. What about soccer? Did he really want to bounce the ball off his head? And so on. In the end, it was clear to him that the inner workings of the human body could not be predicted with certainty and that common sense would have to rule. He couldn't change his life completely, but he could modify the higher-risk activities, and I had no crystal ball.

Disc herniations in the low back and neck usually occur unrelated to any real trauma. They can happen to a person who does heavy lifting at work, a secretary who never lifts more than a stack of charts, or a couch potato who lifts nothing but a fork and a channel changer. It's usually not anyone's fault, although if it just happens to take place during the nine-to-five hours while filing charts at work rather than while picking up a towel off the bathroom floor at home, then the workplace can be blamed. It's a "work-related injury." In such a case, I feel sorry for both the worker and the employer because often, the event that precipitates the actual herniation is really just the straw that broke the camel's back. The disc probably weakened slowly over time, more because of the individual biology and physiology of that person's particular disc as opposed to any specific event at work. Even a simple sneeze can be the final straw. (And where were you when that sneeze occurred?) It's a strange system.

I probably annoyed a lawyer who met me in my office regarding one of her clients. She tried to pin me down: "So, did the car accident cause his back problems or not?" I couldn't say for sure, and had to keep it at that. His back was riddled with arthritis, and that doesn't develop overnight or from a car accident. And, once you have bad arthritis, almost anything can set it off.

The crazy part about this whole cause-and-effect dilemma is that a surgeon's reimbursement for an operation is dramatically higher if the herniation is deemed to be "work related" or "accident related" because a different type of insurance kicks in, as opposed to a patient's regular health insurance. So, for example, if the event occurs at home, the surgeon may be paid less than $2,000 total for the operation and the subsequent three months' worth of postoperative care, phone calls, prescription refills, employer paperwork, and so on. If the event occurs a couple hours later, while the patient is at work, and can be deemed work related, then the surgeon may get closer to $8,000 for the exact same operation and postoperative follow-up.

There are other curious phenomena as well. Various studies have suggested that a patient's actual clinical recovery can be partly tied to whether or not they are receiving disability payments, with payments being associated with slower and less complete recovery. Surgeons talk about this phenomenon among themselves and take it to be self-evident. On the other hand, patients who work for themselves, have a job they love, or have no access to disability payments tend to recover more quickly and more fully.

In working with the spine, I have had an interesting window into a certain facet of human variability. I have met people along the way who, for whatever reason, seem particularly prone to trauma. Clumsiness is a highly variable trait. I once saw a young woman with a ruptured disc in her neck and took a detailed history. I asked her about her pain history, lifestyle, recent falls or accidents. "Well, I've fallen down the basement stairs a few times." A few times? Her husband said she's always been accident-prone. It's strange: some people who are otherwise perfectly healthy are just more prone than others to slip and fall, twist their ankle, hit their head on something, or get rear-ended. It may be that they are less sure-footed, maybe a little less aware of their surroundings, or simply slow to react. Regardless, I'm sure that a clumsiness rating scale will be invented at some point, and I'm sure that health insurers will have a field day with it.

In my practice I'm strongly in favor of giving patients as much information as possible. I like to direct them to the better Web sites, go over their scan results in detail, clue them in to the current internal controversies of the field, and encourage second opinions if they have the inclination. What I'm a little wary of, though, is sending someone a formal report of their MRI scan before I have a chance to go over everything with them. The words in the report can set off such a flurry of anxiety that the subsequent visit requires twice as

long: half the time to explain the results and the other half to manage the anxiety that has built around phrases like "facet hypertro-
phy," "thickened ligamentum flavum," and "incidental note made of
a hemangioma of L2."

Radiologists, in order to cover their own butts in this age of exu-
berant litigation, need to point out each and every tiny finding, regardless of significance, and may even dictate something like
"cannot rule out infection" when infection is really the most remote
possibility and they know it. They have to. And, because a spine naturally ages over time, practically any MRI report on a patient over
age forty or so will be at least slightly "abnormal" in some way. But,
in many cases, certain findings should be no more anxiety provoking
than what might be dictated if a physician were to do a detailed report on a patient's outward appearance: "touch of gray around tem-
ples, bilateral crow's-feet, and incidental note of dry skin on left
cheek." It's just that the words are less familiar in an MRI report.

You can often tell when a patient has read his own report ahead of
time, before the office visit. He may tell you: "My L5 hurts," in direct reference to the level of his back mentioned in the report as having some sort of finding. In general, it's best not to do that. Better to
just say that you have low back pain and to let the physician try to
sort it out. Just because there is a minor abnormality on the scan does
not mean that it is definitely the source of the pain. On the flip side,
some people with terrible back pain have a completely normal MRI.

As another insider's piece of advice, it's best to describe your pain,
or any other symptom, as plainly as possible, like "stabbing pain
down the back of my right leg," rather than dramatizing it as "a
thousand—maybe ten thousand—little fire ants crawling over my
skin, and stabbing me with those little tiny pincers, you know those
little pincer things they have." Doctors can be wary of what they feel
may be overdramatization (fair or not) and may even use the dreaded
phrase "out of proportion"—as in "pain out of proportion to the relatively minor findings on the MRI." It's not that they won't believe

you, it's just that they might be encouraged to end the visit as quickly as possible.

Similarly, if you have a history of a disc herniation (or disc bulge, disc protrusion) in your back, you don't get extra sympathy, from a surgeon at least, if you refer to it as a "disc *explosion*," with wide eyes and dramatic arching arm motions to accompany the term.

I do have tremendous sympathy, though, for any patient trying to make their way through the dizzying array of options when it comes to treating a spine problem. The fact that there are so many options—with new ones always coming on the market and old ones falling out of favor—means that there is really no single best option. You just have to pick one, try it, and move on to the next one if it doesn't do the trick. Sometimes time is all that is needed, as the body often heals itself, but whatever intervention a patient happened to have tried last will be credited as the miracle cure, with "tincture of time" getting none of the credit.

And another thing: as surgeons, we don't get any specific training in the fine distinctions between different mattresses, office chairs, or shoes. This falls out of the realm of science and medicine and you'd probably do just as well to ask the same question of your mother or cousin if they've experimented with any of the above for their own back. If you want a more professional opinion, though, a chiropractor, physical therapist, or even the salesperson at the local "back store" would probably be able to engage in a more informative and enthusiastic discussion. I once had a patient who asked if I wouldn't mind accompanying her out to her car to see if the driver's seat might be contributing to her back problem, and should she get a new car?

New surgical spine procedures are being developed more rapidly than ever before, and each area of the country is covered by a roving team of industry representatives who know all the facts about their particular instrumentation, their brand of screws and rods, and whatever revolutionary device their company has just rolled out. And they're more than happy to take a surgeon out to dinner. The rapid pace of new developments is amazing to see and, in general, I'm very enthusiastic about innovation and the American way, as long as new devices are approached with at least some skepticism at first. (But not too much skepticism. That can be just as bad as none at all.)

What impresses me is the number of patients willing to be guinea pigs: to consider becoming early recipients of a new technology—not just buying the latest version of an iPod, but going through the latest operation or having the latest implant placed in their spine. A current trend, for example, is the idea of disc replacement. There are many ways to handle a worn-out disc: live with it; take pain medication as needed; manage it with continued conservative care like physical therapy, chiropractic manipulation, acupuncture, et cetera; have surgery to remove the disc and fuse the disc space; or try the new disc replacement procedure (which requires surgery through, or just to the side of, the abdomen).

The replacement option makes a heck of a lot of intuitive sense. After all, a disc is a type of joint. When a hip goes bad we don't fuse it, we replace it! This logic, so clear and simple, is certainly alluring to patients. Furthermore, there are data from the enlightened continent of Europe and the procedure is now approved in the United States. So why not? Well, what about the slowly developing information that perhaps 10 percent of disc replacement patients may require repeat surgery at some point, and that repeat surgery can be risky because of scarring around major blood vessels? And how do these mechanical discs look in ten, twenty years?

The bottom line, from my point of view, is that disc replacement technology is probably here to stay, but it may be overused in the heady early years until we figure out which patients are really the most appropriate candidates. So, the earliest patients are performing a valuable service, not only potentially for themselves, but equally for future patients who may either accept or shy away from this option based on their longer-term data. For the time being, I remain an intrigued observer.

———

An aging mind is even more mysterious than an aging back. Dementia, especially, can be a tough problem to tease apart, especially when the patient himself denies the problem and is dragged in by a spouse. When memory starts to fail, most people will jump to one of two conclusions: old age or Alzheimer's disease. In general, it's best not to assume either one at first. If you assume old age, you may decide not to seek help. If you assume Alzheimer's disease, you may be overwhelmed by a sense of futility. A good dementia workup by a neurologist is the way to go, before jumping to any conclusions.

Most forms of dementia, unfortunately, are irreversible. True, there are a couple drugs on the market that can slow down the process, but none can really halt or reverse the deterioration and they never seem to work the miracles that we all hope they will. The problem is too complex to be treated by a single drug. The diagnosis and treatment of dementia is still in its infancy, but I do remain optimistic about the future.

Neurosurgeons are not typically involved in the diagnosis and treatment of dementia, except in the less common *reversible* forms in which surgery is an option. We're not on the front lines of the dementia war, and I don't claim to be the world's expert. What I can say, though, is that helping one of the lucky patients with a reversible form of dementia can be quite satisfying.

Very rarely, a tumor such as a large one pressing on both frontal lobes may be responsible for the cognitive changes. Surgery to remove the tumor can improve the deficits. More commonly, an elderly patient will be diagnosed with an even more mysterious entity: normal pressure hydrocephalus, or NPH. Some small percentage of patients (there are no exact numbers) who are thought to have

Alzheimer's disease actually have NPH, and it takes a savvy internist, neurologist, or television-watching family member to consider the diagnosis in the first place.

I'm serious about the television-watching family member. A layperson, for example, will often ask for certain drugs by trade name or description (like the "purple pill") because of television and magazine ads. Doctors are used to that. What's new is that a patient or family member may ask about a particular device, one used to treat NPH, because of a television ad. I have seen elderly patients coming out of the woodwork recently—usually accompanied by an Internet-savvy daughter or son—with the potential diagnosis of NPH. They see the ad on television, look up the disorder on the Internet, discover that a programmable shunt is the answer, type in their zip code on the device company's Web site, and find out that I'm in their area.

Normal pressure hydrocephalus is an unusual condition. It occurs almost exclusively in the elderly brain. At its core, there is an underlying problem with absorption of cerebrospinal fluid, which not only surrounds the brain, but fills the ventricles, the fluid spaces deep within the brain. There are four ventricles: the two lateral ventricles, the third ventricle, and the fourth ventricle. Normally, cerebrospinal fluid is produced and absorbed at a rate of about 450 cc per day (almost half a liter). At any given time, about 150 cc of cerebrospinal fluid surrounds the nervous system—the typical soda can holds 355 cc—which means that the entire volume is produced and absorbed three times over a twenty-four-hour period. If not absorbed properly, it tends to build up, causing enlargement of the ventricles.

Although the ventricles of an elderly person are naturally larger than a younger person's, due to atrophy, in NPH we say that the ventricular enlargement is "out of proportion to the atrophy." The ventricles are larger than expected for the patient's age. The "normal" in the name of the disease sets it apart from the better-known hydrocephalus of childhood, which causes elevated pressure. So, unlike the typical hydrocephalus, it does not cause headaches and has no potential to be acutely life threatening. Nobody knows why it happens or even exactly why it causes that particular triad of symptoms. It still falls into the "bad luck" category of disease etiology.

In most NPH patients, memory changes are not the only symptom. The other two common symptoms are difficulty in walking (a "magnetic gait" in which the patient tends to shuffle, failing to pick up his feet adequately) and urinary incontinence. Some patients have the entire triad and some have only one of the symptoms. The treatment is the same as that for the childhood form of hydrocephalus: insertion of a shunt that drains fluid directly from the brain into the abdomen.

This has been done for years, but without much fanfare. Now, because of the advertising campaign, the public's interest has been piqued. The wife of one patient told me: "His doctor recommended a shunt a few years ago and we weren't interested, but when I saw that commercial a few weeks ago . . . the man they showed was able to walk so well after surgery!" That's the power of the media.

The shunt procedure, even though it takes only an hour or so, can actually be one of the most satisfying in neurosurgery when it works well, which is most of the time but not all of the time. (One of the most frustrating things is the inability to reliably weed out the patients who won't respond well. Also, sometimes a patient will regress after an initial period of postoperative improvement, which can be equally frustrating.) I tend to remember best the ones that work best.

I had a woman recently who brought her father back to my office three weeks after surgery, beaming with "I have my father back!" Previously, wheelchair bound and with worsening dementia, he had been quite withdrawn. An aide would have to prompt him to call his daughter and, even then, he would speak for a minute or two and then let the receiver drift down into his lap.

Within of week or so of undergoing surgery, he took the initiative himself to call his daughter, and they spoke for an hour. They had a lot to catch up on. Thrilled, I was still just a little puzzled. His mind was coming back so quickly but he was still in a wheelchair. I figured that his muscles must have atrophied and would need to be strengthened first. With confidence, though, I told the patient at the end of his visit: "You're a star patient. You should be up and walking soon."

Two days later I received a message to call the patient's physical therapist. Now what, I thought—did he fall? Do they need some sort of prescription from me? It turns out she called to share the good news that my patient had just gotten up and walked, for the first time in several months, after informing everyone: "The doctor told me I should be able to walk."

It can be quite interesting to watch a mind emerge from a fog. I had a similar patient who had been withdrawn and also quite cantankerous. His family had grown weary of him. They were a bit skeptical of the shunt, but also desperate, so the wife went ahead and signed the consent as power of attorney. The patient had no idea as to what he was about to undergo, or, if he did have some idea, he would forget by the next day.

After surgery, the reemergence of his mind was slow and steady, his walking improved, and he became less grouchy as a bonus. During one of our appointments, after nearly three full months of seeing him back in my office, checking his incisions, talking to him and his family, and evaluating his postoperative scans, he looked at me—really looked at me—and at that very moment seemed to have finally put everything together: the surgery, his improvement, his family's delight. Everything clicked, right then and there in my office.

"Tell me, Doctor," he said. "*You're* the one who put this shunt in me, aren't you?"

I nodded and smiled, amused at his delayed reaction, delayed by nearly three months.

"Well," he added, "you did a fine job."

TWENTY

Brainlifts

People often ask me what the future of neurosurgery will look like. Given our growing understanding of how the brain works, and the constant evolution of technology, picture this potential scenario.

Laura Grasso—the *client*, not patient—is a well-dressed and intelligent woman, a former high-powered lawyer. She took six years off while her kids were young, but she always knew she would return to the practice of law. Now she is ready. She wants her career back. There are only two problems: a couple of her former partners are a little skeptical (behind her back), and she no longer feels on top of her game. She fears what other former professional mothers in her neighborhood also fear in themselves: her mind has turned to mush.

In reality, her mind hasn't quite turned to "mush," as her husband reassures her, but her gut tells her she's not as quick and as sharp as she had been six years ago. Back then—before kids—she was reading voraciously, preparing arguments, and working long hours alongside brilliant colleagues. She had always prided herself on her ability to remain physically fit and slim, even as a busy mother of two, but she hadn't managed to do as much for her mind as she had done for her body. There are only so many hours in the day. Now she needs a cognitive tune-up, especially in the memory department, and she is willing to pay for the best.

Grasso chose her physician, Dr. Lawrence Steele, based on tips from trusted friends and on her recent reading of a glowing *New York Times Magazine* article. Steele is the owner and director of the New York Center for Cognitive Enhancement on Park Avenue, on the Upper East Side of Manhattan. His office is slick, modern, high end. It doesn't look like a doctor's office. He accepts cash only. Insurance doesn't cover what he does because his work does not involve treating illness. This is not medicine in the traditional sense. He is one of the happiest doctors around because his clients are some of the most satisfied high-achievers around.

For years, of course, an entire special breed of physician has used a medical degree to treat normal people, people with no specific illness. Enhancement of otherwise normal healthy people—via facelifts, Botox injections, collagen injections, breast implants, liposuction—has become so commonplace that there are no longer any plastic surgery reality shows on television. The voyeuristic intrigue of plastic surgery had run its course. The public had become so accustomed to seeing both celebrities and the average woman next door before and after their "work" that the transformations were no longer that interesting. They had become entirely predictable. And, the thought of a doctor who didn't treat illness was no longer considered unusual at all. Finally, neurosurgeons and neurologists were able to take advantage of all the benefits of caring for the healthy (or the "worried well") that plastic surgeons had enjoyed for all that time.

In his younger years, Steele was a rising superstar in an academic neurosurgery department. He had specialized in vascular neuro-surgery, for aneurysms of the brain. There weren't many such super-specialists left, as much of the aneurysm surgery had been replaced by minimally invasive treatments performed largely by radiologists.

He started to have second thoughts about his career when his wife forced him to sleep in the guest room during the weeks he was on call. She had grown tired of the rude awakenings from his pager, night after night. She could never fall back asleep after being forced to overhear lengthy phone conversations about the crisis of the mo-ment in the ER or ICU. Sometimes she surprised him with the jargon she would pick up along the way: PComm, AComm, vasospasm, CT angio. In addition to the sleepless nights, she resented how he always seemed to get called away from dinners, movies, and important fam-ily events. In her mind, that was no way to live, and she didn't keep those thoughts to herself.

In public, Steele liked to blame his wife for the change in his ca-reer focus, but privately, he knew that her complaints were only part of the picture. The truth was that he was being paid less and less to do more and more, as reimbursement levels dropped and as fewer neurosurgeons were willing to operate on high-risk, high-litigation cases—like ruptured aneurysms—where the outcomes were often poor. He seemed to specialize in the worst of the worst.

The pride and bravado he had once had for his ability to handle the toughest challenges in the OR had waned. His priorities and de-sires were different now. He became desperate for a change, but didn't think he was prepared to make a radical shift into venture cap-ital or a start-up medical device company as a couple of his col-leagues had done. He would maintain a clinical practice, but a very different type of practice. So he left his academic department and made the move into his own high-end private practice. The move paid off. Now, there was no call schedule, no sleeping in the guest room, and no missed birthdays. His wife was pleased and so was he.

Steele had many clients just like Grasso. He knew he could help her. He listened to her story, took a few notes, and examined the detailed cognitive questionnaire she had filled out while sitting on the Le Corbusier couch in his orchid adorned waiting room.

"So, what can you do for me?" asks Grasso, her Montblanc pen poised above her Italian leather-bound notebook.

"Well, your memory 'challenges'—we don't use the word 'deficit'—would put you into Class Three of our memory scale, a scale that ranges from one to ten. A Class Three challenge is certainly not significant enough to be noticed, say, by other moms in casual conversation at the neighborhood jungle gym, but enough to be noticed by your more perceptive colleagues at a stressful meeting, for example. Stress, of course, can heighten any cognitive challenge."

"That's what I worry about," Grasso admits. "I want to come back strong. I don't want my partners talking behind my back—'she's not the star she used to be,' 'she lost her edge'—I want to blow them away."

"Okay, then, let's figure out how we can blow them away," Steele says, with a laugh. He loves this type of patient. "I have three main options to offer you, of varying levels of intensity and invasiveness."

Grasso's pen gets started.

"The least invasive is a memory training system, computer-based, that can be used right on your own laptop; very discreet. It requires a real time commitment, but it's absolutely risk-free. Some of my clients like to start with this approach and then switch to a more potent option if they're not satisfied with the results. My particular program, designed here at my center, is far superior to the more proletarian programs you may have heard about, the ones popping up at various community centers around the city."

"How long until you see results?" Grasso asks, looking up from her notebook.

"It can be weeks or even up to three or four months, depending on how diligent you are and the hours you're able to put in," Steele explains.

"Hmm . . . doesn't sound right for me. I've got two kids and . . ." Grasso trails off.

"Okay, fine," Steele continues. "Then there's TMS, or transcranial magnetic stimulation, which you probably read about in the *Times* piece."

"Yes, TMS," Grasso confirms.

"The clunky models that were used years ago have been redesigned," Steele explains. "They're also more potent. The demand for TMS is overwhelming, and the demand is what's spurring all the innovation. I'm even brewing a couple ideas of my own. Anyhow, you're not paying me to hear about my patent ideas.

"Compared to the first option, the results are quicker. Because the newest models are able to penetrate to greater depths with greater ease—reaching the hippocampus for memory enhancement, for example—the treatments aren't bad: half an hour, three times a week, typically. And, your time's not wasted. We've developed our own elite spa service that offers manicures, pedicures, and massages while you undergo your TMS treatments, so you can multitask. That's important for someone like you." Steele winks.

Grasso smiles. "So what's the downside?" she asks.

"There's a very small risk of setting off a seizure, but very low," Steele says. "We're talking one in five hundred patients or so with the new model and technique. But, of course, if you happen to be that one in five hundred, you're not so happy about it."

"One in five hundred . . ." Grasso says, looking up toward the ceiling. "That's not a bad statistic."

"No, not at all," Steele agrees. "But before you settle on TMS, let me tell you about the last option. We like to call it a 'brainlift' because it's more dramatic and much longer lasting—hopefully permanent, actually. What Botox is to a facelift, TMS is to a brainlift. Although a brainlift requires an operation—that I perform—it's not major surgery. I usually do these procedures on a Friday and my clients are actually back to work by Monday."

"So. Let me ask you this. I've had my eyes done. How does it compare to that?" Grasso asks.

"Very similar, very similar," Steele explains. "It's about as minor as that, but you could consider it even easier in some respects. Any bruising that occurs is completely invisible. It's very slick. With the hair-parting technique I use, I don't have to shave any hair at all, and the stitches are not visible."

"Well, I don't know if I'm quite willing to consider 'brain surgery,' " Grasso admits, a bit skeptical.

"This is *not* brain surgery, *per se*!" Steele tells her, leaning forward on his desk. "Let me explain what I do here. We map out your memory network based on functional MRI—that's completely painless, of course. Then we take you to the OR and you're put under general anesthesia. I work with only the best anesthesiologists. We make a small incision—half an inch—in the scalp overlying each major node in the memory network, create a small hole in the skull, and insert a neat little metal plug, similar to a watch battery, that contains both a stimulator-electrode and battery. We close everything up with fine absorbable sutures, and that's it. It's nearly impossible for anyone to tell that you even had surgery."

"And how does it work?" Grasso asks, intrigued but still skeptical.

"The stimulation is constant and low grade," Steele explains, "which is superior to TMS, which delivers higher-intensity stimulation, but only intermittently. In a sense, the implants are a more 'natural' method. And about the batteries—which I'm sure is your next question—they are recharged as needed, usually only every two to three years, noninvasively, right through the scalp, right here in the office. There's no need for repeat surgery."

"Well, let me think about it, but I'm leaning toward TMS . . . and the manicure! I don't know if my husband is ready for a bionic wife. I'll let you know."

Grasso closes her notebook, stands up, and shakes hands with Steele, who smiles and predicts another satisfied customer.

Steele returns home to his Upper East Side apartment and notices that his wife has left another article for him on the counter of their newly renovated kitchen, this time from *Atlantic Monthly*. Cognitive enhancement is hot, and everyone wants to know more about it. This article focuses more on the ethical debate as opposed to the science. He's been seeing more and more of this kind of thing recently.

The ethics behind cognitive enhancement is the one deepening wrinkle in this growing trend. Academicians—many of whom have never even spoken to satisfied clients such as his—claim that cognitive enhancement threatens to broaden the socioeconomic gaps in society. The fear is that the wealthy will continue to get ahead, leaving less room at the top for the poor folks who can't afford such procedures. And what will be done about the growing number of parents asking for implants for their underage children? The press is eating this stuff up, taking sides, and exaggerating the controversies.

But Steele maintains a balanced perspective on all of this chatter. Plastic surgery triggered similar debates years ago, but the debates didn't last. Brainlifts will go through the same cycle: they'll gain broader acceptance, the debates will eventually die down, the procedures will become more commonplace. Prices may even drop at some point, making them more affordable, at least to a broader middle class. And why wouldn't more people be willing to pay out of pocket when they see what a brainlift has done for their spouse, a parent, or a friend? After all, we're talking about a more finely tuned mind, not just a tighter face.

On one hand, with the prospects for cognitive enhancement, I think neurosurgery will expand, offering more procedures to a larger swath of the population. On the other hand, I think it could shrink, as disorders that were once subjected to surgery will be treated in whole new ways. Neurosurgery may involve less surgery in the future and maybe, even, fewer surgeons. If we are creative enough, though, we may actively prevent it from shrinking, if we can simply redefine the surgeon.

Already certain entities, like aneurysms of the brain, that used to almost invariably require opening up the head, are now more often being treated by snaking a thin catheter through a vessel in the groin, navigating it up to the aneurysm, and shooting tiny metal coils into it, blocking it off and preventing it from bursting. This doesn't require a surgeon—it can be done by a specially trained radiologist—but some neurosurgeons have redefined themselves to span the role of surgeon and interventional radiologist, undergoing additional training (and sometimes enduring acrimonious turf battles) in order to perform the coiling procedure. The ones who don't redefine themselves in this way, due to indifference, resistance, or maybe advanced age, may be at risk of losing business or eventually becoming extinct.

Tumors may become the next aneurysms. Already, many smaller tumors that used to be subjected to surgery, especially ones in trickier locations, instead undergo the focused radiation treatment of radiosurgery, such as with the Gamma Knife. Although no incision is required, savvy neurosurgeons have held on tightly to this nonsurgical treatment. In reality, the procedure could be performed primarily by radiation oncologists, with surgeons serving more as consultants (this is actually how it is done at some centers). The creator of the technology, though, was an inventive neurosurgeon, so it was born into our specialty and will probably stay there.

Glioblastomas, or GBMs, the malignant tumors that arise from the brain tissue itself—the tumors that have been the target of much research but tragically little enhancement of survival—are still primarily surgical holdouts, even though they cannot be cured by surgery or by anything else. In the future, though, I bet that traditional surgeons will play a more minor role in the treatment of these tumors. Diagnosis may be made solely on the basis of advanced imaging, perhaps combined with a blood or spinal fluid test rather than surgical biopsy, and treatment may be via injection of a gene therapy agent, transported by special viruses programmed to "infect" and attack only tumor cells. Treatment will be tailored to the patient's individual tumor biology (not all tumors are exactly alike), and a perfectly customized chemotherapy cocktail will be mixed for each patient. Designer medicine may well arrive in full force and fulfill its promise.

If this happens, a turf battle may break out between neurosurgeons and oncologists, as I suspect that neurosurgeons—unwilling to lose their hold—will decide to double-train in neuro-oncology and spend more time in the clinic than in the OR. In that event, I wouldn't be surprised if innovative surgeons then invent devices that somehow enhance or deliver the treatment in a novel way, lending a more surgical feel to the process. A few surgeons have already toyed around with early models of drug delivery directly into the brain or cerebrospinal fluid, via a surgically implanted pump, and I'm sure this innovation will, at some point in the near future, make the leap from the experimental to the more routine.

Strokes are far more common than brain tumors. Neurosurgeons, as opposed to neurologists, typically enter the stroke world in only a few circumstances: to clean out clogged carotid arteries in the neck to prevent future strokes, to operate on a less common form of stroke caused by bleeding (as opposed to the far more common "ischemic stroke" caused by blockage of blood flow), and, rarely, to operate on very large ischemic strokes that have swollen and become acutely life-threatening. Other than that, the neurologist is the primary stroke caretaker.

At present, once a stroke occurs, there is little that can be done specifically for that dying or dead portion of brain. The white-hot exception, though—and this is why it is *so* critical to treat stroke as an absolute emergency—is if the stroke is caught within the first three hours of symptom onset. In that case, which doesn't happen nearly enough, a medication (tPA, or tissue plasminogen activator) can be injected in the ER to try to dissolve the clot blocking the vessel. This may restore blood flow and prevent the affected portion of brain from dying.

Unfortunately, most patients present to the ER or their physician after the critical three-hour time window. This happens for many reasons. Some people don't know enough to take their symptoms seriously, symptoms like the sudden onset of slurred speech, or weakness or numbness on one side of their body. Because a stroke is often painless, unlike most heart attacks, a sense of urgency and panic may not set in. I have often heard patients say: "I thought I should just wait to see if it would go away." In essence, what they are really saying, without knowing it, is "I thought I should just wait to see if a part of my brain was going to live or die." Another problem is that some people wake up in the morning with a new stroke symptom, so the time of onset is impossible for anyone to know. And other people, especially who live alone, become impaired to the point where they cannot seek help, and no one else is around to take action on their behalf.

For most stroke victims, then, a portion of brain dies and a neurological deficit sets in. Luckily, the impaired functioning does tend to improve over time, at least to some degree. For example, most people rendered unable to walk by a stroke eventually improve to the point of walking again. The plasticity of the brain, meaning its inherent ability to either move function around or strengthen a previously more minor part of the network, can work wonders. However, recovery can be incomplete, leaving a stroke victim with a significant limp or an arm that is capable of large-scale movement but no fine control or finger dexterity.

An exciting innovation on the near horizon—already in human research trials at the moment—may benefit stroke patients and put neurosurgeons squarely in the middle of the stroke world. Suppose you take a seventy-year-old woman who has a stroke that leaves her unable to move the left side of her body. She presents to the ER six hours after her symptoms start, too late for tPA. A neurologist takes care of her while she's in the hospital. She is stabilized and then sent to a rehabilitation center a week later.

In today's world, that's about all that can be done. We hope that her brain recovers as fully as possible, and that rehab can help support that natural process. Six, eight, or ten months later, she has reached the point of maximal recovery. She still drags her left foot and cannot use her left hand well. She will not continue to improve past that point.

But what if you could scan her brain, figure out what portion of it has become responsible for movement of her left side, and then stimulate that area of cortex, via an electrical implant, to coax it to work even better? That's what the trial is all about. And, assuming success with motor recovery, treatment for post-stroke aphasias, or speech disturbances, would be the next logical step. If all goes well, such a technological breakthrough could ignite a revolution in the treatment of stroke patients. Stay tuned. . . .

———

Here is what I see if I look even further into the future of brain surgery. I may be going out on a limb, but I suspect that when the future arrives at surgical enhancement of the mind, certain individuals will come to desire more than just a cognitive "tune-up." These select few will undergo implantation of electrical stimulation devices in order to develop a savant-like mind. They will become "savant specialists" in a society that already values super-specialization. Some will excel specifically in mathematics, others in linguistics, and some in memory, visual-spatial, musical, or various creative skills. Maybe we will even be able to intensify empathy.

These elite savants will be a boon to the professions and organizations that tap into these particular strengths. They will expand the potential for human achievement and productivity beyond what is currently constrained by average human brainpower.

This isn't possible yet because the riddle of the savant mind hasn't been figured out in enough detail. But I have no doubt that it will, with all the research into how the mind works, or doesn't work. Typically, the major downside of having a rare, naturally occurring, savant mind is that it comes at great cost to other mental abilities. (Recall the offensive-sounding term "idiot savant.") Such a person may be able to multiply 1,456 by 6,321 off the top of his head, or may be able to recite pi out to hundreds of digits, but he may not be able to hold a meaningful conversation with another human being. One part of the savant brain is revved up at the expense of other parts. That much is understood, and that is why most savants would not be able to hold down a job or contribute to society in a significant way.

I would bet, then, that the coaxing of a savantlike mind from an otherwise normal brain would require a two-pronged approach: stimulation of one area of the cortex and simultaneous suppression of another. This would be done in a controlled, scientific way. A patient—or client—could consciously choose to suppress a function they assume they won't miss, like their visual-spatial skills (losing their sense of direction, for example, or their ability to rotate an object in their minds), while revving up their memory to beyond normal human standards. And, because the procedure would be performed only in people who were already mature enough to make such a decision, they would remain capable of the normal human interaction they had already developed through life, allowing them to remain functional—and now even more "valuable"—members of society.

Most people, of course, would want absolutely nothing to do with this. The majority would prefer the concept of a "well-rounded" person, or the ideal of the "Renaissance Man." Some might even feel that the surgically mediated creation of a savant mind smells faintly of eugenics.

I can see, though, how certain individuals could be interested—very interested. Imagine a single man, late twenties, who is anxious to rise up the socioeconomic ladder, quickly. He feels he's at the bottom and has nothing to offer. He has no special skills or natural talents. He can't find a good job. Then this: He reads that an investment of $10,000 (okay if paid in installments) in a new minimally invasive surgical procedure can dramatically enhance the cognitive function of his choosing. He chooses memory. Memory, of course, is intimately connected to learning. With this new, seemingly limitless talent, he decides to learn five new languages (in one year), join the CIA, travel the world, and impress his colleagues by repeating—verbatim—hour-long conversations held with key informants. He has no sentimental feelings for his lost spatial abilities: he uses a Global Positioning System to navigate through unfamiliar cities.

A neurosurgical implant for savant-inducing purposes could be quite powerful, but also difficult to control. What if a newly minted savant uses his cognitive superiority for crime, or terrorism? What if one savant can do the work of a dozen normal people, leading to the unemployment of former colleagues? No doubt, various interest groups would lobby to ban the practice altogether.

Once the technology for cognitive enhancement exists, though, it will be hard to stop. If banned in one country, it will pop up in another. What if the United States outlaws the procedure but China and India embrace it? For every thousand people interested in becoming savants in the United States, there might be tens of thousands in China or India, all striving to climb the same ladder.

The neurosurgeon in me wonders how this implant would affect the individual brain. The anthropologist in me wonders how it might change the world.

Acknowledgments

Much of this book is based on my seven years of residency training, and so I thank *all* of the neurosurgeons who trained me, who taught me to think and act like a neurosurgeon until I became a neurosurgeon. I owe particular thanks to Peter Jannetta, Dade Lunsford, Dennis Spencer, Doug Kondziolka, Donald Marion, Howard Yonas, Leland Albright, Peter Sheptak, and Amin Kassam.

Equally, I thank all the residents who trained alongside me. Each one could have written a similar book, and better. In fact, much of the humor in these pages was inspired by my coresidents, some of whom will recognize phrases that I borrowed (or stole?) from them. They'll have to forgive me. I worried that if I didn't write all this down, it might be lost forever.

In particular, I thank Atul Patel and Kevin Stevenson for the entire seven-year run. And I have to give special thanks in the humor department to John Wahlig, Chris Comey, Brian Subach, and David Lowry. Don't worry, I won't reveal who did or said what.

Thank you to the late Julio Martinez for being the best neuropathologist on earth and for serving as a role model in maintaining a fresh curiosity throughout life.

I am, of course, indebted to the numerous patients I have encountered along the way. I can't thank them enough. They are the whole reason that we do what we do.

I credit Tom Petzinger for getting this ball rolling in the first place, for reading the initial kernel of this book and then passing it on to his literary agent, Alice Martell, who took the ball and ran with it. I will always remember the moment I checked my e-mail at a café in San Diego, while attending a neurosurgery conference, to see a note from Tom telling me that Alice was interested, and wanted to meet me in New York. Thank you, Tom!

And to Alice Martell, my agent, who made everything happen. I really admire her intellect, keen eye, sound judgment, professionalism, and sense of style. The two days I spent with her pitching my idea to publishers in New York opened my eyes to a new world. I thank her for giving me the grand tour and for her continued guidance.

Susanna Porter, my editor at Random House, is top-notch. I trust her instincts. I thank her for making this book much stronger than it was in its fledgling form. If only every new author could be so lucky. Her assistant, Johanna Bowman, was a great help every step along the way as well.

Kirsty Dunseath, my editor at Weidenfeld & Nicolson in the UK, was a pleasure to work with. I was encouraged by her enthusiasm and greatly appreciated her editorial expertise.

I thank Chris Philips, at the American Association of Neurological Surgeons, for her prompt and friendly responses to my questions about neurosurgery census data.

I give my appreciation to my neurosurgery partner, Zoher Ghogawala, for being such an outstanding partner, and I thank him in advance for covering any time I spend away for this book.

Many thanks to my in laws, Russ and Emily Firlik, for their enthusiasm, curiosity, and encouragement along the way.

Tremendous gratitude, of course, goes to my entire wonderful Schreiber family: my parents, Hal and Helen, and my siblings, Ingrid, Richard, and Elizabeth. I thank my family for all their support, always. I'm a lucky kid!

And for Andy, for everything, ever since the day we met.

Notes

Chapter 2: Small World

1. Samuel H. Greenblatt, "The Image of the 'Brain Surgeon' in American Culture: The Influence of Harvey Cushing," *Journal of Neurosurgery* 75 (1991): 808–11.
2. K. S. Firlik and A. D. Firlik, "Harvey Cushing, MD: A Clevelander." *Neurosurgery* 37 (1995): 1178–1186.
3. John F. Fulton, *Harvey Cushing: A Biography* (Springfield, Ill.: Charles C. Thomas, 1946), 713–14.

Chapter 7: Evolution Through Blood

1. More recently, the concept of a "minimally conscious state" has also emerged, which does complicate things a bit. See Joseph T. Giacino, et al., "The Minimally Conscious State: Definition and Diagnostic Criteria," *Neurology* 58 (2002): 349–53.
2. Francis Crick and Christof Koch, "A Framework for Consciousness," *Nature Neuroscience* 6 (2003): 119–26; Margaret Wertheim, "After the Double Helix: Unraveling the Mysteries of the State of Being," *The New York Times*, Apr. 13, 2004, F3.

3. Francis Crick and Christof Koch, "What Is the Function of the Claustrum?" *Philosophical Transactions of the Royal Society B: Biological Sciences* 360 (2005): 1271–79.

4. Wertheim, "After the Double Helix."

5. Francis Crick, et al., "Consciousness and Neurosurgery," *Neurosurgery* 55 (2004): 273–82.

Chapter 9: Risk

1. Douglas Kondziolka, Mark R. McLaughlin, and John R. Kestle, "Simple Risk Predictions for Arteriovenous Malformation Hemorrhage," *Neurosurgery* 37 (1995): 851–55.

Chapter 11: Disturbing Deviations

1. R. J. Haier, et al., "Structural Brain Variation and General Intelligence," *Neuroimage* 23 (2004): 425–33.

2. Sophia Frangou, Xavier Chitins, and Steven C. R. Williams, "Mapping IQ and Gray Matter Density in Healthy Young People," *Neuroimage* 23 (2004): 800–805.

3. K. H. Lee, et al., "Neural Correlates of Superior Intelligence: Stronger Recruitment of Posterior Parietal Cortex," *Neuroimage* (in press).

4. Sharlene D. Newman and Marcel Adam Just, "The Neural Bases of Intelligence: A Perspective Based on Functional Neuroimaging," in Robert J. Sternberg and Jean Pretz, eds., *Cognition and Intelligence: Identifying the Mechanisms of the Mind* (New York: Cambridge University Press, 2005), 88–103.

Chapter 12: Slices

1. K. S. Firlik, A. J. Martinez, and L. D. Lunsford, "Use of Cytological Preparations for the Intraoperative Diagnosis of Stereotactically Obtained Brain Biopsies: A 19-year Experience and Survey of Neuropathologists." *Journal of Neurosurgery* 19 (1999): 454–58.

2. Nitzan Mekel-Bobrov, et al., "Ongoing Adaptive Evolution of ASPM, a Brain Size Determinant in Homo Sapiens," *Science* 309 (2005): 1720–22; Nicholas Wade, "Brain May Still Be Evolving, Studies Hint," *The New York Times*, Sept. 9, 2005.

Chapter 15: Traces of Thought

1. Marcel Adam Just, et al., "Brain Activation Modulated by Sentence Comprehension," *Science* 274 (1996): 114–16.

2. N. Sadato, et al., "Activation of the Primary Visual Cortex by Braille Reading in Blind Subjects," *Nature* 380 (1996): 526–28.

3. Leonardo G. Cohen, et al., "Functional Relevance of Cross-Modal Plasticity in Blind Humans," *Nature* 389 (1997): 180–83.

4. Lawrence Osborne, "Savant for a Day," *The New York Times Magazine*, June 22, 2003, 38.

5. A. Sterr, et al., "Changed Perceptions in Braille Readers," *Nature* 391 (1998): 134–35.

6. Daniel Goleman, "The Lama in the Lab," *Shambala Sun*, Mar. 2003, 64–70; Daniel Goleman, "Finding Happiness: Cajole Your Brain to Lean to the Left," *The New York Times*, Feb. 4, 2003, F5.

KATRINA FIRLIK was the first woman admitted to the neurosurgery residency program at the University of Pittsburgh Medical Center, the largest—and one of the most prestigious—neurosurgery programs in the country. She is now a private practitioner in Greenwich, Connecticut, and a clinical assistant professor at Yale University School of Medicine. She lives in New Canaan, Connecticut, with her husband, a neurosurgeon turned venture capitalist.

ABOUT THE TYPE

This book was set in a digital version of Monotype Walbaum. The original typeface was created by Justus Erich
Walbaum (1768–1839) in 1810. Before becoming a punch
cutter with his own type foundries in Goslar and Weimar,
he was apprenticed to a confectioner where he is said to
have taught himself engraving, making his own cookie
molds using tools made from sword blades. The letter-
forms were modeled on the "modern" cuts being made at
the time by Giambattista Bodoni and the Didot family.